HEUER CHRONOGRAPHS

The Jack Heuer Years

1962–1982

ARNO MICHAEL HASLINGER

photography by Clemens Kois

Leading Timekeepers
in Motorsport

Air, Sea, Military,
and Olympic Timekeepers

Jack Heuer's Years

Timekeeping Formula 1

HEUER CHRONO GRAPHS

CALLWEY

HEUER
CARRERA
McLAREN
WORLD CHAMPIONS
1974
SWISS
21
TEXACO
TEXACO-Marlboro
HEUER
Lockheed
CHAMPION
TEXACO

INHALT/
CONTENTS

Emerson Fittipaldi gewann 1974 die Formel 1 Weltmeisterschaft für McLaren. Tag Heuer ist heute der offizielle Zeitmesser von McLaren Mercedes.
Emerson Fittipaldia won the Formula 1 Championship for McLaren in 1974. Tag Heuer is the official Timerkeeper of McLaren Mercedes today.

VORWORT/ *FOREWORD*

Die Leidenschaft für HEUER-Uhren und den Motorsport ist eine faszinierende Kombination, die Enthusiasten auf der ganzen Welt in ihren Bann zieht. Die Marke HEUER, heute unter dem Namen TAG Heuer bekannt, hat sich nicht nur als fester Bestandteil der Uhrmacherkunst etabliert, sondern auch eine untrennbare Verbindung zur Welt des Motorsports aufgebaut.
HEUER-Uhren sind seit 15 Jahren ein fixer Bestandteil der internationalen Auktionswelt und erzielen regelmäßig beeindruckende Ergebnisse. Während einige Modelle Ikonenstatus erreicht haben und in Sammlerkreisen höchst begehrt sind, warten andere noch darauf, entdeckt zu werden. Die anhaltende Sammelleidenschaft für HEUER-Uhren zeigt, dass diese Zeitmesser weit mehr sind als bloße Werkzeuge zur Zeitanzeige. Sie verkörpern ein Lebensgefühl und haben individuelles Design, Geschichte sowie technische Raffinesse.
Dieses Buch beschreibt eine besondere Ära zwischen 1962 und 1982, in der Jack W. Heuer in vierter Generation die Firma leitete. Als Marketingspezialist war es faszinierend zu beobachten, wie es ihm gelang, die Marke HEUER in der Welt des Motorsports zu etablieren. Durch die Partnerschaft mit Ferrari, von 1971 bis 1979, beschritt Heuer neue Wege der Produktplatzierung. Das weiße Marken-Logo schien (als erster Sponsor der Uhrenindustrie) von der Front der roten Ferraris und war plötzlich auf allen Formel-1-Grand-Prix-Strecken der Welt omnipräsent. Die goldenen Heuer Carreras, sowie Modellinien wie Camaro, Jarama, Silverstone, Monza und Modena sind eng mit dieser Periode verbunden.
Die Einzigartigkeit des rechteckigen Designs der Heuer Monaco war so avantgardistisch, dass Steve McQueen entschied, die Monaco für seinen Film „Le Mans“ zu tragen. McQueen wollte im Film so authentisch wie möglich sein und wählte einen Rennanzug wie sein Freund Jo Siffert, der Markenbotschafter von HEUER war. So kam der Heuer-Aufnäher in den Film Le Mans und die Monaco auf den rechten Arm von Steve McQueen. Diese revolutionären Marketingideen von Jack Heuer sind auch Thema dieses Buches.
Legendäre Rennfahrer wie Mario Andretti, Tony Adamowicz, Derek Bell, Emerson Fittipaldi, Mike Hail-

A passion for HEUER watches and motorsport is a fascinating combination that captivates enthusiasts all over the world. The HEUER brand, now known as TAG Heuer, has not only established itself as an integral part of the art of watchmaking, but has also forged an inseparable link with the world of motorsport.
HEUER watches have been a fixture in the international auction world for 15 years and regularly achieve impressive results. While some models have reached iconic status and are highly sought after in collectors' circles, others are still waiting to be discovered. The enduring passion for collecting HEUER watches shows that these timepieces are far more than mere tools for telling the time, symbolizing an attitude to life, style, history, and technical sophistication.
This book describes a special era between 1962 and 1982, when Jack W. Heuer was the fourth generation to run the company. As a marketing specialist, he has been fascinating to observe as he succeeded in establishing the HEUER brand in the world of motorsport. Partnering with Ferrari from 1971 to 1979, Heuer broke new ground in product placement as the first in the Swiss watch industry to sponsor Formula 1, a white HEUER logo emblazoned from the front of the red Ferraris, omnipresent on Formula 1 Grand Prix circuits worldwide. The golden HEUER Carreras and model lines such as Camaro, Jarama, Silverstone, Monza, and Modena are closely associated with this period.
The uniqueness of the rectangular design for the Heuer Monaco was so avant-garde that Steve McQueen decided to wear the Monaco for his film "Le Mans." He wanted to be as authentic as possible and chose racing overalls like his friend Jo Siffert, who was a HEUER brand ambassador. That's how the Heuer patch came to be on the chest of Steve McQueen and the Monaco on his right wrist. These pioneering marketing strategies by Jack Heuer are also the subject of this book.
Legendary racing drivers such as Mario Andretti, Tony Adamowicz, Derek Bell, Emerson Fittipaldi, Mike Hailwood, James Hunt, Graham Hill, Jacky Ickx, Niki Lauda, Arturo Merzario, Ronnie Peterson, Clay Regazzoni, Carlos Reutemann, Jochen Rindt, Jody Scheckter, Rolf Stommelen, Gilles Villeneuve, Björn Waldegaard,

wood, James Hunt, Graham Hill, Jacky Ickx, Niki Lauda, Arturo Merzario, Ronnie Peterson, Clay Regazzoni, Carlos Reutemann, Jochen Rindt, Jody Scheckter, Rolf Stommelen, Gilles Villeneuve, Björn Waldegaard und andere vertrauten auf HEUER-Uhren während der wilden 1970er-Jahre, den sogenannten „Heyday" des Motorsports.
Seit den 1960er-Jahren erweiterte Jack Heuer das Uhrenportfolio kontinuierlich. Neue Modelllinien wie die Heuer Cortina, Daytona, Jarama, Kentucky, Modena, Montreal, Monza, Silverstone und Verona wurden nach berühmten Städten oder Rennstrecken benannt. Andere Modelle, wie Heuer Calculator oder Skipper zeichneten sich durch ihre funktionale Namensgebung aus. Diese besondere Vielfalt macht Heuer-Chronographen heute spannend für Sammler.
TAG Heuer hat die Geschichte der Marke geschickt gepflegt und weiterentwickelt. Die Wiederbelebung der ikonischen Markenwerte ist beeindruckend und verleiht HEUER-Uhren heute neuen Glanz. Seit 2016 ist TAG Heuer offizieller Zeitnehmer und Partner des Red Bull-Formel 1- Weltmeisterteams.
Eine Sensation ist der Einstieg von LVMH als Hauptsponsor der Formel 1, die damit 2025 Rolex ablöst. Die Partnerschaft, die auf zehn Jahre angesetzt ist, könnte TAG Heuer viele neue Möglichkeiten eröffnen.
Im September 2024 übernahm Antoine Pin die Rolle des CEO bei TAG Heuer. Mit seiner Markenerfahrung und der Unterstützung von Jean Christophe Babin als CEO der LVMH-Uhrensparte ist dies eine äußerst vielversprechende Kombination.
In dieser erweiterten Ausgabe von Heuer Chronographen gibt es neue Kapitel zu entdecken, die einen noch tieferen Einblick in die Welt von HEUER gewähren. Dieses Buch bietet Sammlern, Experten sowie Neueinsteigern gleichermaßen wertvolle Informationen und Tipps zur besseren Identifikation authentischer Uhren.
Ein besonderes Highlight sind die zahlreichen neuen Uhrenfotos von Clemens Kois, die es ermöglichen, Modelle in ihren faszinierenden Facetten zu bewundern. Detailreiche Aufnahmen zeigen charakteristische Merkmale und die einzigartige Patina, die diese Sammlerstücke im Laufe der Jahre entwickelt haben. Diese Fotografien sind nicht nur ein optischer Genuss, sondern auch ein wertvolles Nachschlagewerk für alle, die sich intensiver mit den verschiedenen Modellreihen befassen möchten.
Ich lade Sie ein, in die faszinierende Welt von HEUER einzutauchen und die einzigartige Verbindung zwischen Präzision, Stil und Motorsport zu entdecken. Lassen Sie sich inspirieren von der Geschichte dieser legendären Marke und vertiefen Sie Ihr Wissen über eine der spannendsten Epochen der Uhrmacherkunst.
Viel Freude beim Lesen!

Arno Michael Haslinger

and others relied on HEUER watches during the wild 1970s, known as the heyday of motorsport.
Since the 1960s, Jack Heuer has continuously expanded the portfolio. New model lines like Cortina, Daytona, Jarama, Kentucky, Modena, Montreal, Monza, Silverstone, and Verona were named after famous cities or racetracks. Others, such as the Heuer Calculator or Skipper, are typified by their functional names. This special variety makes HEUER chronographs so exciting for collectors today.
TAG Heuer has skillfully cultivated and developed the history of the brand. The revival of the iconic brand values is impressive and gives HEUER watches new luster today.
Since 2016, TAG Heuer has been the official timekeeper and partner of the Red Bull Formula 1 World Champions Team.
2025 saw the sensational entry of LVMH as the main sponsor of Formula 1 2025, replacing Rolex. The partnership is set to run for ten years and could open new opportunities for TAG Heuer.
In September 2024, Antoine Pin took over the role of CEO at TAG Heuer. His brand experience and the support of Jean Christophe Babin as CEO of the LVMH watch division represent a very promising combination.
In this expanded Heuer Chronographs edition, there are new chapters that provide an even deeper insight into the world of HEUER. This book offers collectors, experts, and newcomers alike valuable information and tips for identifying authentic models.
As a highlight, numerous new watch photos by Clemens Kois make it possible to admire all the captivating details of each model. Detailed shots show characteristic features and the unique patina that these collector's items have developed over the years. These photographs are not only a visual delight, but also a valuable reference work for anyone wishing to take a closer look at the various model series.
I invite you to immerse yourself in the fascinating world of HEUER and discover the unique combination of precision, style and motorsport. Let yourself be inspired by the history of this legendary brand, and deepen your knowledge of one of the most exciting eras in the art of watchmaking.
Happy reading!

Arno Michael Haslinger

EINLEITUNG/ *INTRODUCTION*

Jack W. Heuer: Interview

1. Entstehung und Expansion des Portofolios (Masterplan und Mythos)

Als ich im Jahre 1958 als junger Diplomingenieur in das Unternehmen eintrat, war eine meiner ersten Aufgaben, eine Standortbestimmung unserer Position im Markt vorzunehmen. Die Uhrenindustrie stand in diesen Jahren in der Schweiz in einem extremen Preiskampf, da damals der Kaufpreis für Standard-Automatikuhren das ausschlaggebende Kaufargument war. Marketing und seriöse Markenpolitik waren mit wenigen Ausnahmen noch total unterentwickelt. Meine Marktanalyse zeigte auf, dass wir im Segment der Borduhren für Rallyes Weltmarktführer waren (65% der Teilnehmer der Rallye Monte Carlo im Jahre 1958 hatten unsere Borduhren). Sie zeigte ebenfalls, dass wir bei Handstoppuhren eine Führungsposition einnahmen und im Segment der Chronographen zu den wenigen Produzenten von qualitativ hochwertigen Chronographen gehörten. Bei den normalen Automatik-Armbanduhren hatten wir keine Chance, mit den Marktpreisen mitzuhalten. Dadurch ergab sich ein Masterplan zur Fokussierung des Ausbaus des Marktanteils im Segment der Stoppuhren (Verdoppelung durch Übernahme unseres Konkurrenten Leonidas) und der Versuch der Innovationen innerhalb des Marktes für Armbandchronographen. Als Resultat dieser Studie kann eine Reihe von mechanischen Chronographen mit neuem Design genannt werden, z.B. Carrera und Autavia. Da aber der Chronographen-Markt infolge der damals fehlenden Automatik sehr stark rückläufig war, waren wir, um zu überleben, sozusagen gezwungen, in die Entwicklung eines Automatik-Chronographen zu investieren und die Lancierung der Automatikmodelle Carrera, Autavia und Monaco im Jahre 1969 voranzutreiben.

2. Herkunft und Bedeutung der Nomenklatur der Namen

Die Marke hatte in allen vier Familiengenerationen stets eine enge Verbindung zur Welt des Sports, und

1. Creation and Development of the Portfolio (Master Plan and Myth)

As a young engineer joining the watch company in 1958, my first task was to evaluate our position in the marketplace. In those years, the watch industry was involved in an extreme price war in Switzerland because the asking price for a standard automatic wristwatch back then was the decisive argument for driving sales. Marketing and serious policies for establishing brand name recognition were, with few exceptions, completely underdeveloped.

My market analysis showed that we were the leading global brand in the market for on-board rally chronographs (65% of participants in the Monte Carlo rally in 1958 used our products). It also showed that we held a leading position in the hand-held stopwatch segment and that we were one of the few producers of high quality chronographs. In the case of normal automatic wristwatches, we couldn't compete with prices in that segment of the market. This led to a master plan in which we focussed on promoting the stopwatch segment of the market (the takeover of our competitor Leonidas led to doubled capacity at this time) and in which we would attempt to innovate within the wrist-chronograph market segment.

The result of this study was the development of a series of mechanical chronographs with new designs, among them the Carrera and Autavia. But due to the fact that the chronograph market was in a slump because of the lack of automatic technology,, we were, in order to survive, forced in a sense to invest in the development of an automatic chronograph and to drive the launch of the automatic models Carrera, Autavia, and Monaco in the year 1969.

2. Origin and Meaning of the Names

The brand, through all four generations of the family, has always had a close connection to the sports world; curiously enough, from the second generation, the successful models have always been given a name. I con-

Jack W. Heuer und Jo Siffert zeigen 1969 stolz den frisch unterzeichneten Sponsorvertrag.
Jack W. Heuer and Jo Siffert proudly present their freshly signed sponsoring contract in 1969.

Niki Lauda und Clay Regazzoni unterzeichnen den Sponsorvertrag als neue Ferrari-Werksfahrer für die Saison 1974.
Niki Lauda and Clay Regazzoni sign the sponsoring contract as new Ferrari drivers for the 1974 season.

eigenartigerweise wurden seit der zweiten Generation die erfolgreichen Modelle stets mit einem eigenen Namen versehen. Ich habe diese Tradition weitergeführt und jeweils gut klingende Ortsnamen mit sportlichem und chicem Image gewählt. Interessanterweise war die Hinterlegung eines solchen Ortsnamens zusammen mit unserer Hauptmarke Heuer problemlos zu schützen. Cortina wurde gewählt wegen der dort stattgefundenen Winter-Olympiade, gleiches gilt für Montreal usw.

3. Architekten des kreativen Designs und der Technik

Für die Armbandchronographen arbeiteten wir damals traditionsgemäß mit den in-house Designern der Gehäuse- und Zifferblattfabrikanten zusammen. Bei den Stoppuhren, wo wir Weltmarktführer waren und bei denen sich die elektronische Revolution der digitalen Stoppuhren weiter durchzusetzen began, waren wir viel innovativer. Wir kontaktierten wahrscheinlich als erste Uhrenfirma einen weltberühmten Industriedesigner in Mailand mit Namen Richard Sapper. Sapper machte das Design von drei Microsplits, welche seinerzeit alle drei in New York im MOMA („Museum of Modern Art“) ausgestellt waren. Wir sind leider nicht im Besitz von irgendwelchen Entwürfen von Richard Sapper.

4. Stückzahlideen und limitierte Auflagen

Wir haben keine Unterlagen über Stückzahlen und aus meiner Erinnerung kann ich nur ungenaue Angaben machen. Pro Chronographen-Modell in Stahl wurden jeweils je nach Modell pro Jahr zwischen 200 und 2.000 Stück verkauft. „Limited Edition“ ist eine neuere Erscheinung und war damals noch kein Marketingwerkzeug! Es kam auch mehrmals vor, dass ein Rennstall wie Carroll Shelby oder Ferrari bei uns Standardmodelle einkaufte und mit ihrem Markenzeichen versehen ließen, wobei die Stückzahlen jeweils zwischen 20 und 100 Stück lagen und als Firmengeschenke verwendet wurden.

5. Motorsport Sponsoring

In den Jahren 1962 bis 1965 haben wir jeweils dem SCCA („Sports Car Club of America“) Zeitnahmechronographen geliehen, um in Sebring beim 12-Stunden-Rennen, welches vom SCCA organisiert wurde, die Zeitnahme sicherzustellen. Dieses war der Beginn unseres Vorgehens, Zeitmessungsinstrumente und -materia-

tinued this tradition and always chose the names of locations that sounded good and had a sporty and chic image. Interestingly, it proved to be a simple matter to register the name of a location in conjunction with our main brand name of Heuer. Cortina was chosen due to the Winter Olympics that were held there, as was also the case for Montreal.

3. Architects of the Creative Design and Technology

For the wristwatch chronographs, we worked together with the traditional in-house designers of the case and watchface production companies. For the stopwatches, where we led the global market, and where the electronic revolution of the digital stopwatch continued to gain momentum, we were much more innovative. We were probably the first watch company to contact a world-famous industrial designer, namely Richard Sapper in Milan. Sapper designed three microsplits, all of which were displayed at the New York Museum of Modern Art in his lifetime. Unfortunately, we do not have Richard Sapper's conceptual designs in our possession.

4. The Unit Numbering Concept and Limited Editions

We do not have any records of the number of units sold, and I can only provide a rough count from memory. Each steel chronograph model sold somewhere between 200 and 2000 units each year, depending on the individual model. "Limited edition" productions are a new development, and back then were not yet used as vehicles for marketing. It was frequently the case that a racing team, such as Carroll Shelby or Ferrari, would buy a standard model from us and have their logo added to the chronograph; in such cases the number of units ranged from approximately 20 to 100 and were mostly given away as corporate gifts.

5. Motorsport Sponsorship

From 1962 to 1965 we loaned the SCCA (Sports Car Club of America) timekeeping chronographs to track time during the 12-hour races in Sebring, which had been organized by the SCCA. This was the beginning of our practice of providing our timekeeping instruments and materials in return for the placement of our

Pour Jack Heuer,
très amicalement

Niki Lauda und Jack Heuer bei einem Treffen 2010 in Wien.

Niki Lauda meets Jack Heuer in Vienna, 2010.

Jack Heuer im Gespräch mit Arno Haslinger.

Jack Heuer talks to Arno Haslinger.

lien zur Verfügung zu stellen und im Gegenzug im Rahmen eines Sponsorings mit unserem Markenzeichen im Rahmen der Veranstaltung respektive auf dem Rennwagen platziert zu sein. Beim Einstieg in die Formel 1 passierte das Gleiche. McLaren und andere Rennställe wurden unterstützt. Aber der große Durchbruch war die technische Kooperation mit der Scuderia Ferrari in den Jahren 1971 bis 1979. Im Vertrag mit Ferrari mussten wir anspruchsvolle Zeitnahmeaufgaben erfüllen und als Gegenleistung musste Ferrari auf allen Rennwagen (Formel 1 und Langstrecke) unser Markenzeichen gut sichtbar auf der Front der Fahrzeuge platzieren. Zusätzlich war vertraglich vereinbart, dass jeder Formel-1-Werksfahrer automatisch von uns finanziell unterstützt wurde, mit der Verpflichtung für den Fahrer, das Markenzeichen auf Schulterhöhe auf dem Rennoverall zu tragen. Mit anderen Worten, ausnahmslos alle Formel-1-Piloten der Scuderia Ferrari der Jahre 1971 bis 1979 trugen unser Markenzeichen.

6. Filmprojekt Le Mans

Wir arbeiteten schon einige Jahre mit einem so genannten „Property Master“ in Hollywood zusammen. Diese Leute, kurz Propmasters genannt, bildeten ein eigenes Syndikat und lieferten an den Filmsets alles, was nicht zu den Kostümen oder dem Bühnenaufbau gehörte, wie z.B. die Coca-Cola-Flaschen oder die Camel-Zigaretten und insbesondere auch die Uhren. So haben uns z.B. Jack Lemmon, Charlton Heston oder Burt Reynolds signierte Bilder geschickt und als Gegenleistung die im Film getragenen Heuer-Chronographen von mir als Geschenk erhalten. Während der Dreharbeiten für den typischen Rennfahrerfilm „Le Mans“ haben wir dem gleichen Propmaster nicht nur ein paar Uhren nach Le Mans geschickt, sondern eine ganze Kiste mit Zeitnahmebrettern, Taschenchronographen, Stoppuhren, Markenzeichen, Aufklebern usw. Viele Leute auf dem Set trugen somit unsere Uhren, wie in der Realität auch, da innerhalb der Motorsportteams unsere Uhren sehr oft getragen wurden.

7. Welche bekannten Persönlichkeiten trugen Heuer-Uhren? Gab es auch Frauen?

Diese Liste von berühmten Namen ist sehr lang. Wenn ich mich auf meine Jahre beschränke, gehören dazu die Könige von Spanien, Griechenland und Thailand, der Prinzgemahl Philip von England und seine Schwägerin

logo as a form of sponsoring on the race cars. When we entered the Formula One market, the same thing happened. We supported McLaren and other racing teams. But the greatest breakthrough came with our technical partnership with Scuderia Ferrari from 1971 to 1979. Contracted by Ferrari, we had to achieve a meticulous level of timekeeping accuracy, and in return Ferrari had to place our logo in an easily visible location on the front of their race cars (both Formula One and Endurance). In addition we were contracted to automatically recompense every Formula One driver in return for them wearing the Heuer logo at shoulder height on their racing suits. In other words, every single Formula One race car driver of Scuderia Ferrari, without exception, wore our logo from 1971–1979.

6. The Le Mans Film Project

We have been working for a number of years with a "property master" in Hollywood. These individuals, "propmasters" for short, maintain their own syndicate that delivers everything to film sets except for costumes and scenery elements, such as Coca-Cola bottles or Camel cigarettes, and in particular also the clocks and watches.

This resulted in Jack Lemmon, Charlton Heston, and Burt Reynolds, among others, sending in signed photographs for one of the Heuer watches they wore in their films. In the case of "Le Mans," as a typical motorsport film, we sent the propmaster in Le Mans not only a few watches, but an entire case with timekeeping boards, pocket chronographs, numerous stopwatches, Heuer patches and stickers, etc. Any people on the set thus wore our watches, just as in real life, where our watches were often worn by the motor sports teams.

7. Which famous personalities wore Heuer watches? Were there any women?

The list of famous names to have worn Heuer watches is very long. If I restrict myself to my time of active service, we can include the Kings of Spain, Greece, and Thailand, Prince Philip of the United Kingdom and his daughter Princess Anne (military stopwatch), and Robert Kennedy in the USA, to name just a few of the most important personalities.

969

Jack W. Heuer begrüßt Enzo Ferrari zum Mittagstisch vor dem Grand-Prix-Start.

Jack W. Heuer greets Enzo Ferrari during lunch prior to the start of the Grand Prix.

Prinzessin Ann (Military-Stoppuhr), Robert Kennedy in den USA, um nur einige der wichtigsten zu nennen.

8. Heuer-Uhren in militärischen Anwendungen
Als einer der führenden Hersteller von Stoppuhren und Chronographen gehörten verschiedene Armee-Organisationen nach dem Ersten Weltkrieg zu unserer regelmäßigen Kundschaft. Diese Liste umfasste u.a.: die königliche britische Marine und Armee, die kaiserliche japanische Marine, die U.S. Marine, Armee und Luftwaffe, die italienische und argentinische Luftwaffe sowie die Schweizer Armee.

9. Meine Lieblingsuhren aus dem Portofolio
Dazu gehört ein 18 Karat Gold Automatik Carrera Chronograph aus den 70er Jahren. Diese gleiche Uhr überreichte ich jedem Formel-1-Fahrer der Scuderia Ferrari während unserer Partnerschaft mit der Scuderia. Die Uhren hatten den Namen und die Blutgruppe des jeweiligen Fahrers auf dem Gehäuseboden eingraviert. Da wir mehrere Fahrer durch Rennunfälle verloren haben, sind diese Uhren für mich mit vielen Emotionen verbunden. Im Weiteren hänge ich sehr an einem limitierten Modell des Carrera Chronographen, welches zu seinem 40-jährigen Jubiläum im Jahre 2004 neu aufgelegt wurde. Dieses Modell trägt meine eingravierte Unterschrift auf dem Gehäuseboden. Diesen Chronographen mit der Nummer 0001 trage ich sehr häufig und auch sehr gerne. Bei den neueren Modellen begeistert mich am meisten die Carrera Calibre 360 in 18 Karat Rotgold.

10. Persönliche Kontakte zu Fahrern und Teamchefs
Ich lernte alle Forme-1-Fahrer, die wir unter Vertrag hatten, persönlich kennen, als ich ihnen ihre Uhren überreichte. Mit einigen hatte ich ein freundschaftliches Verhältnis, mit anderen weniger. Die Teamchefs kannte ich alle und mit dem Commendatore Ferrari hatte ich ein sehr gutes Verhältnis.

Jack W. Heuer

8. Heuer chronographs in military applications
As one of the leading producers of stopwatches and chronographs, our customer base following the First World War included various military organizations, among them the British Army and Navy, the Imperial Japanese Navy, and U.S. Marines, the Army, and Air Force, the Italian and Argentinian Air Forces and the Swiss Army.

9. My favorite watches from the portfolio
One of my favorite watches is the 18 karat-gold-automatic Carrera chronograph from the 1970s. I presented this watch to every Formula One driver of Scuderia Ferrari during our partnership. The watches had the names and the blood types of the individual drivers engraved on the casing. These watches have a deep emotional meaning for me, because we lost a number of drivers to racing accidents. I am also very fond of the limited edition of the Carrera chronograph model that was reproduced in 2004 to celebrate the 40th anniversary of the Carrera line. This model carries my signature engraved on the casing. I often wear this chronograph, engraved with the number 0001, and very much enjoy doing so. Among the newer models, I am excited about the Carrera Calibre 360 in red 18-karat gold.

10. Personal contact with driver and team captains
I met all of the Formula One drivers we had under contract when I personally presented them with their Heuer wristwatch. I had a friendly relationship with some of the drivers, knew all of the team captains, and had a very positive rapport with the Commendatore of Ferrari.

Jack W. Heuer

SPORTS TIMING
HEUER
HEUER
HEUER

Heuer Zeitmessung und Niki Lauda, der den Ferrari 312T bei Regen testet, 1974.

Heuer timing devices and Niki Lauda test driving the Ferrari 312T in the rain, 1974.

Jonathan Darracot

Die Sicht eines Auktionators
Im Jahr 2010 war ich bereits seit über 15 Jahren im Auktionsgeschäft für Uhren tätig, als wir von einer Versteigerung bei Bonhams hörten: Heuer-Uhren aus der Sammlung von Arno Haslinger. Die Idee einer Auktion mit Uhren eines einzigen Sammlers war revolutionär. Die erfolgreiche Auktion veränderte die Landschaft des Heuer-Sammelns, und bis heute ist der Katalog sehr begehrt. Als ich bei Bonhams anfing, wurde ich Arno Haslinger, unserem europäischen Berater, vorgestellt. Sofort verstand ich die Leidenschaft und das Wissen, die es ihm ermöglicht hatten, die erste Heuer-Auktion zusammenzustellen.
In der Folge hat Bonhams zwei bedeutende Sammlungen verkauft, das ursprüngliche Interesse daran wird auf die Teilnahme und Käufe bei der Haslinger-Aktion zurückgeführt. Als wir die erste Sammlung sahen, war eine Heuer Skipper mit dem Spitznamen „Skippererra" das Highlight. Sie wurde für 81.250 Pfund verkauft und stellte damit einen Weltrekordpreis auf – wie auch andere Uhren dieser Auktion.
Bonhams Cars hat einige der besten Autos der Welt verkauft und verfügt über enge Verbindungen zu den bedeutendsten Namen der Rennsport-Community. Dadurch konnte Bonhams Watches einige der großartigsten Heuer Carrera (Referenz 1158) Chronographen versteigern, die von den legendärsten Namen der Formel 1 getragen wurden. Dieser Aspekt hat die goldene Heuer Carrera zu einem der bedeutendsten Chronographen der 1970er-Jahre gemacht. Bemerkenswerte Carrera-Exemplare, die von Bonhams verkauft wurden, gehörten dem Motorradweltmeister Mike Hailwood und Ferrari-Pilot Arturo Merzario – beide mit persönlichen Widmungen von Jack Heuer.
All dies kann den Verkaufsergebnissen von Bonhams entnommen werden, die ein umfassendes Verzeichnis der Heuer-Auktionen seit der Haslinger Collection enthalten – darunter auch die TAG Heuer von Ayrton Senna, die er 1993 seinem Mechaniker schenkte, sowie eine Kopie von Ronnie Petersons 1158, die von TAG Heuer gefertigt und 2018 zugunsten der Ronnie-Peterson-Stiftung beim Monaco Historic Grand Prix mit einem erstaunlichen Ergebnis versteigert wurde.
Es war und ist mir eine Freude, durch die Auktionen von Bonhams an der fortwährenden Geschichte von Vintage-Heuer Uhren teilzuhaben.

Jonathan Darracott
Global Head of Watches
Bonhams Auctions

An auctioneer's view
In 2010—I had already been in the watch auction business for over 15 years—we heard about a sale at Bonhams of Heuer watches curated from the collection of Arno Haslinger, the idea of a single owner Heuer sale was revolutionary. The successful auction changed the landscape of Heuer collecting and to this day, the catalogue is highly sought after. When I started at Bonhams I was introduced to Arno Haslinger, our European consultant. I immediately understood his passion and knowledge that had enabled him to curate the first Heuer sale.
Consequently, Bonhams have sold two impressive collections that have attributed their original interest to attending and buying in the Haslinger sale. When we saw the first collection, a Heuer Skipper nicknamed the "Skippererra" was the highlight of the collection. It went on to sell for £81,250, setting a world record price at the time, as did others in the sale.
Bonhams Cars has sold some of the world's finest cars and has deep connections with the seminal names in the racing community, resulting in Bonhams Watches selling some of the greatest Heuer Carrera (reference 1158) chronographs worn by the most iconic names in Formula 1. This connection has elevated the golden Heuer Carrera to become one of the elite group of chronographs from the 1970s. Notable Carrera examples sold by Bonhams belonged to motorcycle champion Mike Hailwood and Ferrari driver Arturo Merzario, both with personal inscriptions from Jack Heuer.
All these details can be found in the Bonhams sales results which gives a comprehensive listing of Heuer watch results since the Haslinger Collection sale, including Ayrton Senna's TAG Heuer gifted to his mechanic in 1993 and a copy of Ronnie Peterson's 1158 made by TAG Heuer, which sold for an astonishing result for the benefit of the Ronnie Peterson Foundation at the Monaco Historic Grand Prix in 2018.
It has been a joy being part of the continuing vintage Heuer Journey through Bonhams auctions.

Jonathan Darracott
Global Head of Watches
Bonhams Auctions

Heuer Skipper

Mike Hailwood's
Heuer Carrera 1158

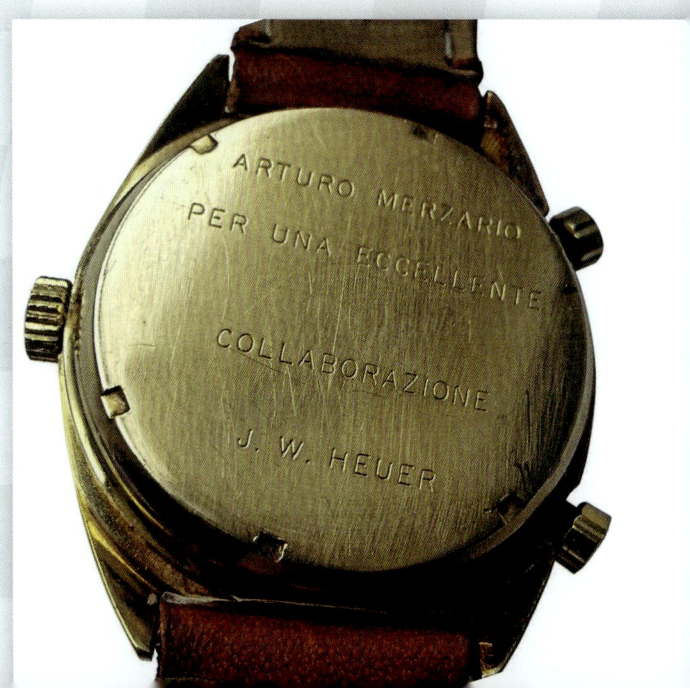

Arturo Merzario's
Heuer Carrera 1158

Heuer Carrera mit Shelby-Cobra-Logo auf dem Zifferblatt.

Heuer Carrera with Shelby Cobra dial

Paul David Maudsley

Meine Beziehung zu Arno Haslinger und Heuer-Uhren war ein wertvoller Aspekt meiner Karriere. Als Uhrenspezialist, Enthusiast und Sammler habe ich schon immer andere Menschen bewundert, die sich mit Leidenschaft und Hingabe dem Sammeln widmen. Arno ist eine solche Person – ein echter Pionier unter den Sammlern von Heuer-Chronographen und jemand, dessen Vision das Sammeln von Vintage-Heuer-Uhren maßgeblich geprägt hat.

Unsere Wege kreuzten sich erstmals 1999, als Arno zu einer meiner Auktionsbesichtigungen bei Bonhams in London kam und sein scharfes Auge sofort auf eine ungetragene, aber alte Heuer fiel. Von diesem Moment an trafen wir uns regelmäßig, tauschten uns über Details aus und ich half ihm dabei, einige der schönsten Heuer-Uhren zusammenzutragen.

Dies führte im Jahr 2010 zur heute geradezu legendären Auktion „Bonhams Haslinger Collection" in London, bei der Arno Bonhams und mir seine unglaubliche Heuer-Sammlung anvertraute. Es war ein Privileg, eng mit jemandem zusammenzuarbeiten, der 10 Jahre damit verbracht hatte, ein so außergewöhnliches Sortiment an Stücken zusammenzustellen. Von Modellen mit Kultstatus wie der Monaco und der Carrera bis hin zu seltenen Prototypen und Unikaten – Arnos Sammlung erzählt die Geschichte von Heuers enger Verbundenheit mit dem Motorsport und Innovation aus Überzeugung.

Arnos Fachwissen, das er in seinem ersten Buch „Heuer Chronographs" zusammengefasst hat, verlieh dem Namen Heuer Renommee und den Lesern historische Einblicke. Gemeinsam präsentierten wir seine Kollektion nicht nur als eine Reihe von Uhren, sondern als kulturelles Zeugnis, das das Ansehen der Marke in der Sammlergemeinschaft erhöhte.

Heute, nach 25 Jahren Freundschaft und noch längerer Sammeltätigkeit, freue ich mich, einen kleinen Beitrag zu dieser wunderbaren neuen Publikation leisten zu dürfen.

Paul David Maudsley
Remontoire Limited

My relationship with Arno Haslinger and Heuer watches has been such a rewarding aspect of my career. As a watch specialist, enthusiast, and collector, I have always admired other individuals who bring passion and dedication to the side of collecting. Arno is one such individual—a true pioneer in the appreciation of Heuer chronographs, and someone whose vision has shaped the landscape of vintage Heuer watch collecting.

Our paths first crossed in 1999 when Arno came to one of my auction viewings at Bonhams London, his keen eye drawn to a new old-stock Heuer. From that moment onwards we met, shared details, and I assisted him in collating some of the finest Heuers ever seen.

This led up to what is now known as the landmark Bonhams Haslinger Collection auction in London in 2010, where Arno entrusted me and Bonhams with his incredible Heuer collection. It was a privilege to work closely with someone who had spent 10 years curating such an extraordinary assortment of pieces. From iconic models like the Monaco and Carrera to rare prototypes and one-offs, Arno's collection told the story of Heuer's deep ties to motorsport and innovation.

Arno's expertise, encapsulated in his first book "Heuer Chronographs," brought credibility and historical insight to the Heuer name. Together, we presented his collection not just as a series of watches, but as a cultural narrative, elevating the brand's standing in the collecting community.

Now after 25 years of friendship and even more collecting, I'm delighted to be able to play a small part in this wonderful new publication.

Paul David Maudsley
Remontoire Limited

phs
HEUER
16
CARRERA
HEUER
60
40
20
45
30
15
SHELBY
COBRA

CHRONOGRAPHEN

Modell-Linien

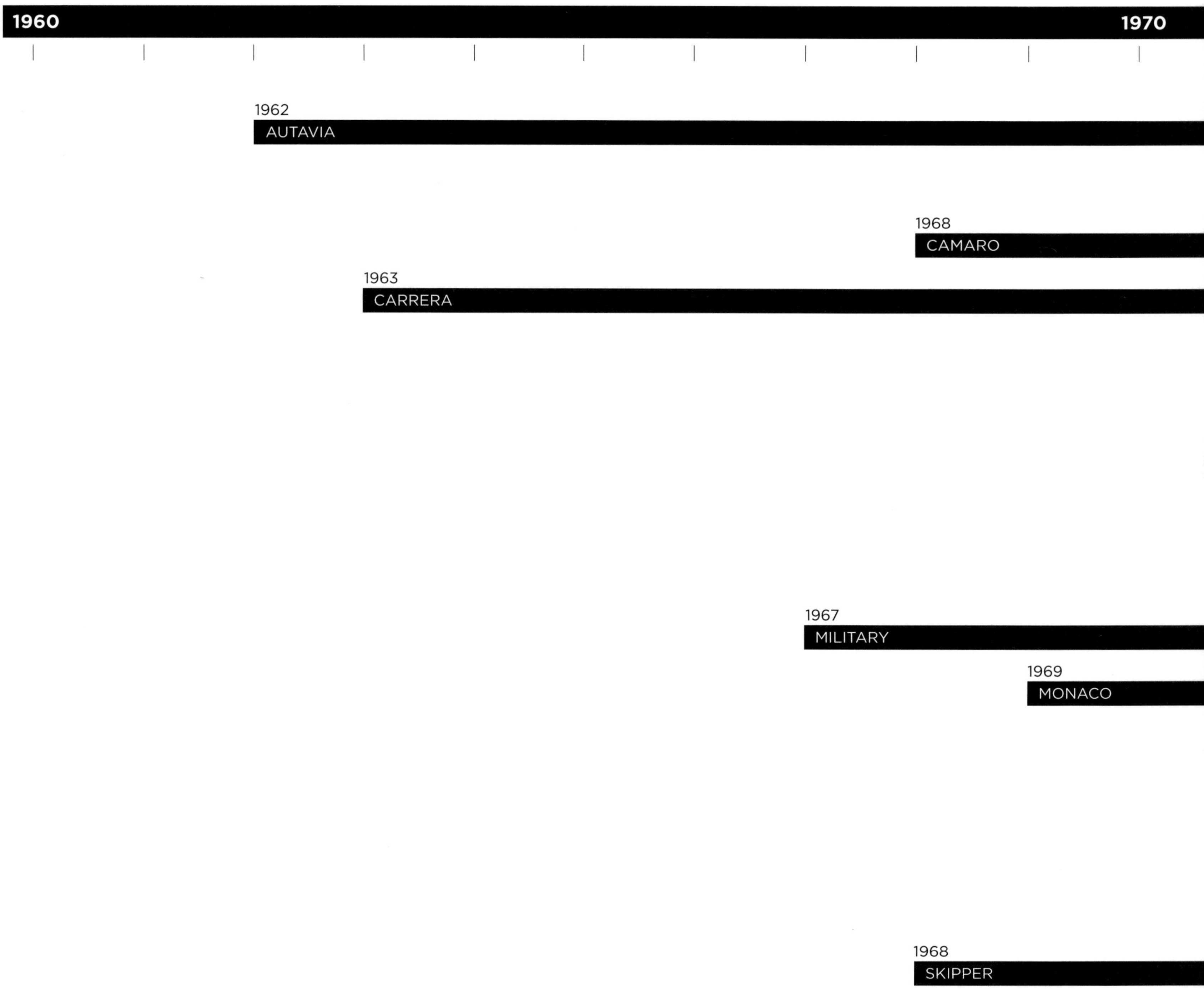

CHRONOGRAPHS

Model Timeline

54A TARGA FLORIO
CIRCUITO DELLE MADONIE
LA CLASSIFICA:
1. SIFFERT/REDMAN PORSCHE-GULF 908-3
2. RODRIGUEZ/KINNUNEN PORSCHE-GULF 908-3
3. Vaccarella/Giunti Ferrari 512 S
4. LAINE/v.LENNEP PORSCHE 908-2
5. WALDEGAARD/ATTWOOD PORSCHE-GULF 908-3
6. Müller/Parkes Ferrari 512 S
PORSCHE
Gulf/Firestone/Bosch

AUTAVIA

Im Hause Heuer hatte man 1962 eine Idee: Sie lancierten einen Chronographen mit „Autavia" Schriftzug über dem Heuer-Logo auf dem Zifferblatt.
Das war der Beginn der Heuer-Modellreihen. „Autavia" steht für AUTo-AVIAtion und verbindet die zwei Geschäftsbereiche: Herstellung von Zeitmess- und Cockpit-Instrumenten für Motorsport und Luftfahrt. „Chronograph" heißt „Zeitschreiber" (griech. chronos „Zeit" und graphein „schreiben").
Die Autavia war ein Novum. Stoßsicher und wasserdicht bis 100 Meter Tiefe. Die drehbare Lünette mit 60-Minuten- oder 12-Stunden-Einteilung qualifizierte sie für besondere Einsätze. Deshalb empfahl der Katalog die Uhr Piloten, Sportlern, Tauchern und Forschern. Das klare, schwarze Zifferblatt-Design mit kontrastierenden Hilfszifferblättern war an Ablesbarkeit kaum zu übertreffen. Es wurden Varianten mit zwei oder drei Hilfszifferblättern angeboten.
Eine Autavia GMT ergänzte die wachsende, immer umfangreichere Kollektion. Die 24-Stunden-Lünette mit zusätzlichem rotem Stundenzeiger machte das Messen einer zweiten Zeitzone möglich, ohne die Chronographen-Funktion zu beeinträchtigen.
Jack Heuer hatte einen Traum: der erste Automatik Chronograph sollte seinen Namen tragen. Er traf Willy Breitling, Hans Kocher von der Firma Büren und die Konstrukteure von Dubois-Depraz, die er für die Entwicklung des Automatikwerks und Chronographen-Moduls gewinnen konnte. Es gelang ihm, eine erfolgreiche Kooperation zu starten, der sich die amerikanische Hamilton Watch Company anschloss, die sich bei Büren eingekauft hatte. Erste Prototypen liefen erfolgsversprechend. Das so genannte Kaliber 11 „Chronomatic", eine Namensverbindung aus „Chronograph" und „Automatic", übertraf alle Erwartungen. Ziel war, das Wettrennen gegen Zenith mit El Primero zu gewinnen.
Am 3. März 1969 war es so weit. Nach vierjähriger Entwicklungsarbeit luden die Partner die Presse nach

In 1962, a new idea came to Heuer: They launched a new chronograph and set the model name "Autavia" above the Heuer logo on the dial.This was the start of the Heuer model series.
"Autavia" comes from AUTo-AVIAtion and combines the two sport areas of Heuer's business enterprise: the production of timekeeping devices for motorsport, and the manufacture of cockpit instruments for civil and military aviation."Chronograph" literally means "time writer," from the Greek chronos (time) and graphein (to write).The Autavia was a fascinating new product, shockproof and waterproof at a depth of up to 100 meters. The rotating bezel with a 60-minute or 12-hour register qualified the watch for unconventional tasks. For this reason, it was recommended for pilots, race car drivers, divers, and researchers. The wonderfully, clear black dial design with contrasting subregisters provided virtually unmatched legibility. Two dials were available, providing either two or three subregisters.
Soon a model incorporating GMT (Greenwich Mean Time) enriched the increasingly diverse Autavia collection. A 24-hour rotating bezel with a new red hour hand made it possible to measure time in a second time zone without impacting on the chronograph's functionality.
Jack Heuer had a dream: The first automatic chronograph was to carry his name. He met with his competitors, Willy Breitling, Hans Kocher von Büren, and the construction engineers of Dubios-Depraz and succeeded in convincing them to participate in a joint development project for automation components and chronograph modules. This was the start of a successful partnership that was later joined by US firm Hamilton Watch Company, which had acquired a major stake in the Büren Watch Company.
The initial prototypes promised success. The Calibre 11 Chronomatic, named from the combination of the terms "chronograph" and "automatic," exceeded all expectations. The declared objective of this partnership then focussed on being the first to present this unique prod-

Genf und New York, um die neue Produktlinie zu präsentieren. Es wurde ein Riesenerfolg! Die Prototypen der ersten Serie nannte man „Chronomatic“ und platzierte „Autavia“ im unteren Drittel des Zifferblatts.
Jack Heuer fand neue direkte Sponsormöglichkeiten: schon immer den Motorsport im Visier, gewann er den Schweizer Rennfahrer und Porsche-Händler Jo Siffert, der fortan eine weiße Autavia der ersten Serie trug, was Fotos eindrucksvoll belegen. Jack Heuer kaufte einen Porsche 911, zusätzlich bekam Jo Siffert noch 25.000 Schweizer Franken für das Tragen des Heuer-Logos auf seinem Rennanzug! Das Heuer-Zeichen tauchte immer öfter auf den werblich wenig genutzten Rennanzügen auf. Willy Breitling fragte Jack Heuer, ob er Chronomatic für seine Uhrenlinie benützen dürfte, was Heuer bejahte, daher verschwand der Schriftzug „Chronomatic“ und wurde durch „Autavia“ ersetzt. 1972 machte Heuer eine Promotionaktion mit dem Viceroy Racing Team in den USA. Beim Kauf einer Viceroy Zigarettenstange konnte man sehr günstig eine Heuer Autavia der Referenz 1163 V mit schwarzem Zifferblatt erwerben.V stand für Viceroy und war zwischen den Bandanstößen eingeprägt. Die Nachfrage war groß, da die Uhr im Handel fast dreimal mal so teuer war und sich nur durch den Tachymeter anstatt der Minuten-Stunden-Lünette unterschied. 1163 V Gehäuse wurden in großer Stückzahl bestellt und fanden auch bei anderen Autavia Varianten dieser Zeit Verwendung.
Das Modell Autavia traf Design-Vorstellungen der Konsumenten in den 70er Jahren. Das vergrößerte Gehäuse verlieh eine Präsenz, der sich niemand entziehen konnte und prägte dadurch den neuen Trend zu größeren Uhren. Eine große, kreative Modellvielfalt zeichnet die Produktstrategie von Heuer aus, was die Marke so einzigartig macht.

uct to the market and win the race against time against the Zenith El Primero.
It was launched on March 3rd, 1969. After four years of development, the partners invited the excited members of the international media to Geneva and New York for a presentation of the new product line. It was an enormous success! The prototypes from the first production run were named "Chronomatic Heuer" and displayed "Autavia" on the lower third of the dial just above the date window.
Jack Heuer sought new possibilities for direct sponsorship. He had always had motorsport in mind. He succeeded in convincing Swiss race driver and Porsche dealer Jo Siffert to wear a first-series Autavia with a white dial, which is impressively demonstrated in numerous photographs. This was a promising start! Jack Heuer bought a new Porsche 911, and in addition Jo Siffert received 25.000 Swiss Francs for displaying the Heuer logo on his race suit. The Heuer logo was now seen over and over again on the suits of the drivers, which had hitherto not been used for advertising. Willy Breitling asked if he could use the term "Chronomatic" for his new model line; Jack Heuer agreed and subsequently removed the word "Chronomatic" while repositioning the model name "Autavia" above the TAG Heuer logo. In 1972 Heuer held a huge promotional event with the Viceroy Racing Team in the US. Purchasers of a carton of Viceroy cigarettes could send in the end flap to obtain a Heuer Autavia reference 1163 V with black dial for one-third of the standard price. 'V' was stamped between the lugs and stood for "Viceroy." The demand was enormous. The Viceroy model had a tachymeter instead of the minutes-hour bezel. 1163 V cases were ordered in great numbers and also used for other Autavia models during this time period.
The Autavia fitted perfectly into the consumer fashion trends of the 1970s. The enlarged case displayed mesmerizing presence and significantly influenced the trend of larger watches. A great variety of models were invented by the creative designers to fulfill the demands of the market and embody Heuer's product strategy, which makes the company unique in its range and variety of products.

Armband-Chrono-graphen

Zifferblätter mit Spezialteilung
Die Chronographen Ref. 404/05, Ref. 2447/48 u. Ref. 3647/48 können mit folgenden Spezial-Zifferblättern geliefert werden:
Schwarz,
Tachymeter (Geschwindigkeitsmesser)
Pulsometer (Pulsmesser)
1/100 Min.-(Dezimal-)Messer

Armband-Gold-Chronographen
in 18 Kt., siehe Preisliste

Spezialprospekt verlangen.

1/5 Sek.-Messer, 0–30 Min.-Zähler. Präzisionswerk, Anker 17 St., antimagnetisch, INCABLOC, unzerbrechliche Feder.

Ref. 404
Stahl, wasserdicht, ⌀ 34,5 mm.

Ref. 405
Doublé, 20 Mikr., Stahlboden, wasserdicht, ⌀ 34,5 mm.

1/5 Sek.-Messer, 0–30 Min.- und 0-12 Stunden-Zähler. Präzisionswerk, Anker 17 St., antimagnetisch, INCABLOC, unzerbrechliche Feder.

Ref. 2447 CARRERA 12
Stahl, wasserdicht, ⌀ 35,5 mm.

Ref. 2448 CARRERA 12
Doublé, 20 Mikr., Stahlboden, wasserdicht, ⌀ 35,5 mm.

1/5 Sek.-Messer, 0–30 Min.- und 0–12 Stunden-Zähler. Präzisionswerk, Anker 17 St., antimagnetisch, INCABLOC, unzerbrechliche Feder.

Ref. 2446 M AUTAVIA 12
Stahl, wasserdicht, Glasring mit 60 Min.-Teilung, ⌀ 38 mm.

Ref. 2446 H AUTAVIA 12
wie oben, jedoch Glasring mit 12 Stundenteilung.

1/5 Sek.-Messer, 0–45 Min.-Zähler
Anker 17 St., antimagnetisch, INCABLOC, unzerbrechliche Feder.

Ref. 3641
Nickel-Chrom, Stahlboden, wasserdicht, ⌀ 35 mm.

Ref. 3645
Doublé, 20 Mikr., Stahlboden, wasserdicht, ⌀ 35 mm.

1/5 Sek.-Messer, 0–45 Min.-Zähler. Anker, 17 St., antimagnetisch, INCABLOC, unzerbrechliche Feder.

Ref. 3647 CARRERA 45
Stahl, wasserdicht, ⌀ 35,5 mm.

Ref. 3648 CARRERA 45
Doublé, 20 Mikr., Stahlboden, wasserdicht, ⌀ 35,5 mm.

1/5 Sek.-Messer, 0–45 Min.-Zähler. Anker, 17 St., antimagnetisch, INCABLOC, unzerbrechliche Feder.

Ref. 3646 H AUTAVIA 45
Stahl, wasserdicht, Glasring mit 12 Stundenteilung, ⌀ 38 mm.

Ref. 3646 M AUTAVIA 45
wie oben, jedoch Glasring mit 60 Min.-Teilung, ⌀ 38 mm.

AUTAVIA
HEUER
SWISS

AUTAVIA

ref. 2446 M „Big Subs“

Erstes Gehäuse mit verschraubtem Boden; schwarzes Zifferblatt und große weiße Hilfszifferblätter. Die 30-Minuten- und 12-Stunden-Totalisatoren befinden sich bei 3 bzw. 6 Uhr. Die permanente Sekunde als Hilfszifferblatt befindet sich bei 9 Uhr. Alle Hilfszifferblätter sind in das Zifferblatt eingelassen, mit vertikalen Seiten und konzentrischen, kreisförmigen Rillen.
Die 12-Uhr-Position ist mit arabischen Ziffern beschriftet, alle Stundenindizes mit Ausnahme der 11- und 1-Uhr-Position sind leicht abgeschnitten, um die großen Hilfszifferblätter zu integrieren. Der Autavia-Schriftzug befindet sich über dem Heuer-Logo. In die Dauphine-Zeiger der zweiten Ausführung aus Stahl ist Leuchtmasse integriert. Der Chronographen-Sekundenzeiger ist weiß lackiert, alle kleineren Zeiger auf den drei Hilfszifferblättern sind schwarz. „SWISS“ ist oberhalb von 6 Uhr aufgedruckt.
Die schwarze Lünette ist in beide Richtungen drehbar, hat ein vergilbtes Leuchtdreieck bei 12 Uhr und zeigt Minutenmarkierungen. Bei dieser frühen Referenz ist die gewölbte Krone unsigniert und von zwei kleinen Chronographen-Drückern umgeben. Der äußere Gehäuseboden ist mit „AUTAVIA HEUER“ und „STAINLESS STEEL CHRONOGRAPH-WATERPROOF GUARANTEED 330FT UNDER WATER“ gekennzeichnet. Die Innenseite des Gehäusebodens ist mit „ED. HEUER & Co SWISS“ und die Werkbrücke mit “ED. HEUER & CO SA UNADJUSTED“ signiert. Dies war der eingetragene Name des Unternehmens vor der Fusion mit Leonidas im Jahr 1964, nach der es offiziell zu Heuer-Leonidas SA wurde. Die Seriennummer befindet sich zwischen den Anstößen bei 6 Uhr. Angetrieben wird sie von einem frühen Valjoux 72-Uhrwerk, das unter der Unruh mit 72 R signiert ist.

First Autavia case with screw back. Black dial and large white subdials. The 30-minute and 12-hour totalizers are positioned at 3 and 6 o'clock respectively. The permanentsecond subdial is at 9 o'clock. All subdials are recessed in the dial with vertical sides and concentric circular ridges.
12 o'clock is written on the dial in Arabic numerals; all hour indices except for 11 and 1 o'clock are slightly cropped to incorporate the large subdials. The Autavia signet is positioned over the Heuer logo. The steel dauphine hands of the second run have integral luminous material. The chronograph second hand is painted white; all smaller hands on the three subdials are black. "SWISS" is printed above 6 o'clock.
The black bezel is bidirectional, with a yellowed luminous triangle insert at 12 o'clock showing minute markings. For this early reference, the domed crown is unsigned and flanked by two small chronograph pushers. The outer case back is marked "AUTAVIA HEUER" and "STAINLESS STEEL CHRONOGRAPH-WATERPROOF GUARANTEED 330FT UNDER WATER." The interior of the case back is signed "ED. HEUER & Co SWISS" and the movement bridge with "ED. HEUER & CO SA UNADJUSTED." This was the company's registered name prior to the 1964 merger with Leonidas, after which it officially became Heuer-Leonidas SA. The serial number is between the lugs at 6 o'clock. Powered by an early Valjoux 72 movement, it is signed "72 R" under the balance wheel.

AUTAVIA

ref. 2446 M

Gehäuse mit verschraubtem Boden. Das Zifferblatt des zweiten Autavia-Modells ist schwarz mit reduzierten weißen Hilfszifferblättern. Die 30-Minuten- und 12-Stunden-Totalisatoren befinden sich bei 3 bzw. 6 Uhr. Das permanente Sekunden-Hilfszifferblatt befindet sich bei 9 Uhr. Die in eine Ebene eingelassenen Hilfszifferblätter haben konzentrische Rillen.
Ein großer Stahlindex befindet sich bei 12 Uhr, reduzierte Stundenstahlindizes sind von 1 bis 11 angebracht, außer bei 3, 6 und 9 Uhr. Runde Leuchtpunkte markieren jede Stunde auf der 1/5-Sekunden-Skala am Zifferblattrand. Der Autavia-Schriftzug befindet sich über dem Heuer-Logo.
Die spitzen Stahlzeiger sind mit Leuchteinlagen versehen. Der Chronographen-Sekundenzeiger ist weiß lackiert, die kleinen Zeiger auf den drei Hilfszifferblättern sind schwarz. Über der 6-Uhr-Position ist „SWISS“ aufgedruckt. Die verblichene bräunliche Minuten-Lünette ist in beide Richtungen drehbar, mit gealterter Leuchtstoffeinlage bei 12 Uhr. Bei dieser frühen Referenz ist die Krone zwischen den Chronographen-Drückern unsigniert. Ins Äußere des Gehäusebodens ist „AUTAVIA HEUER“ und „STAINLESS STEEL CHRONOGRAPH-WATERPROOF GUARANTEED 330FT UNDER WATER“ graviert.
Das Innere des Gehäusebodens ist mit „ED. HEUER & Co SWISS“ signiert, die Werkbrücke trägt bereits die Signatur „HEUER-LEONIDAS SA“, ein gutes Beispiel für die Übergangsphase der Nomenklatur. Angetrieben wird es von einem frühen Valjoux 72, das unter der Unruh mit 72 R signiert ist.

Case with screw-back construction screw back. This second Autavia model has a black dial with reduced white subdials. The 30-minute and 12-hour totalizers are positioned at 3 and 6 o'clock respectively. The permanent second subdial is positioned at 9 o'clock. The subdials are recessed in the dial with sloping sides and concentric circular ridges.
A large steel marker is at 12 o'clock, with plain steel hour markers except at 3, 6, and 9 o'clock. Luminous points mark every hour on the 1/5-second scale at the dial edge. The Autavia name is positioned over the Heuer logo.
Pointed steel hands have luminous inlays. The chronograph second h-and is painted white, the small hands on the three subdials are black. "SWISS" is printed above 6 o'clock. The faded brownish minute bezel is bidirectional, with an aged luminous insert at 12 o'clock. For this early reference model, the crown is unsigned between the chronograph pushers. The outer case back is signed "AUTAVIA HEUER" and "STAINLESS STEEL CHRONOGRAPH-WATERPROOF GUARANTEED 330FT UNDER WATER."
The interior of the case back is signed "ED. HEUER & Co SWISS"; the movement bridge already bears a "HEUER-LEONIDAS SA" stamp, a good example of a transitional mixed nomenclature. Powered by an early Valjoux 72 movement, it is signed 72 R under the balance wheel.

AUTAVIA
HEUER
SWISS

Jochen Rindt im Lotus 72 bei der Startaufstellung für den Österreich Grand Prix 1970.

Jochen Rindt in his Lotus 72 on the grid of the Austrian Grand Prix in 1970.

Jochen Rindt trägt eine HEUER Autavia 2446 M.

Jochen Rindt wearing a HEUER Autavia 2446 M

AUTAVIA

ref. 2446 T „Jochen Rindt“

Autavia-Gehäuse mit verschraubtem Gehäuseboden. Ein schwarzes Zifferblatt mit 3 weißen Hilfszifferblättern. Die 30-Minuten- und 12-Stunden-Totalisatoren befinden sich bei 3 bzw. 6 Uhr. Die permanente Sekunde als Hilfszifferblatt befindet sich bei 9 Uhr. Die Hilfszifferblätter sind vertieft, mit abgeschrägten Seiten und konzentrischen kreisförmigen Rillen.
Ein großer Stahlindex befindet sich bei 12 Uhr, außer bei 3, 6 und 9 Uhr sind reduzierte Stundenstahlindizes von 1 bis 11 angebracht. Quadratische Leuchtpunkte befinden sich an der Außenseite der Indizes. Der Autavia-Schriftzug ist über dem Heuer-Logo positioniert.
Die spitzen Stabzeiger, die im Durchmesser mit den Indizes übereinstimmen, sind mit Leuchteinlagen versehen. Der Chronographen-Sekundenzeiger ist weiß lackiert, die kleinen Zeiger auf den drei Hilfszifferblättern sind schwarz.
„SWISS“ und „T“ für Tritium sind auf dem Stundenzifferblatt über 6 Uhr aufgedruckt.
Die schwarze Tachymeter-Lünette ist in beide Richtungen drehbar.
Der äußere Gehäuseboden ist mit „AUTAVIA HEUER“ und „STAINLESS STEEL CHRONOGRAPH-WATERPROOF GUARANTEED 330FT UNDER WATER“ signiert.
Das Innere des Gehäusebodens und die Werkbrücke sind mit „HEUER-LEONIDAS SA“ signiert. Angetrieben wird sie von einem Valjoux 72, das unter der Unruh mit 72 R gekennzeichnet ist.

Autavia case with screw back. A black dial with 3 white subdials. The 30-minute and 12-hour totalizers are positioned at 3 and 6 o'clock respectively. The permanent second subdial is positioned at 9 o'clock. The subdials are recessed with sloping sides and concentric circular ridges. A large steel marker is at 12 o'clock, with plain steel hour markers from 1 to 11, except at 3, 6, and 9 o'clock. Luminous points highlight the outer side of the markers. The Autavia name is positioned over the Heuer logo. Pointed baton hands of the same diameter as the hour markers have luminous inlays. The chronograph second hand is painted white, the small hands on the three subdials are black. "SWISS" and "T" for tritium is printed on the hour subdial above 6 o'clock. The black tachymeter bezel is bidirectional. The outer case back is signed "AUTAVIA HEUER" and "STAINLESS STEEL CHRONOGRAPH-WATERPROOF GUARANTEED 330FT UNDER WATER."
The interior of the case back and movement bridge are stamped "HEUER-LEONIDAS SA." Powered by a Valjoux 72 movement, it is signed 72 R under the balance wheel.

AUTAVIA
HEUER
TACHYMETER
SWISS

AUTAVIA

ref. 2446 M „Full Lume“

Autavia-Gehäuse mit verschraubtem Gehäuseboden. Ein schwarzes Zifferblatt mit 3 weißen Hilfszifferblättern. Die 30-Minuten- und 12-Stunden-Totalisatoren befinden sich bei 3 bzw. 6 Uhr. Die permanente Sekunde als Hilfszifferblatt befindet sich bei 9 Uhr. Die Hilfszifferblätter sind vertieft, mit abgeschrägten Seiten und konzentrischen kreisförmigen Rillen.
Ein doppelter Leuchtindex befindet sich bei 12 Uhr, einzelne Leuchtindizes sind von 1 bis 11 angebracht, bei 3, 6 und 9 Uhr befinden sich kleine Leuchtbalken.
Der Autavia-Schriftzug ist über dem Heuer-Logo angebracht. Die spitzen Stabzeiger, die im Durchmesser mit den Indizes übereinstimmen, sind mit Leuchteinlagen versehen. Der Chronographen-Sekundenzeiger ist weiß lackiert, die kleinen Zeiger auf den drei Hilfszifferblättern sind schwarz. „SWISS“ und „T“ für Tritium sind auf dem Stundenzifferblatt über 6 Uhr aufgedruckt.
Die schwarze Minuten-Lünette ist in beide Richtungen drehbar. Eine große Krone ist mit Heuer signiert. Der äußere Gehäuseboden ist mit „AUTAVIA HEUER“ und „STAINLESS STEEL CHRONOGRAPH-WATERPROOF GUARANTEED 330FT UNDER WATER“ gekennzeichnet. Das Innere des Gehäusebodens und die Werkbrücke sind mit „HEUER-LEONIDAS SA“ signiert. Angetrieben wird sie von einem Valjoux 72, das unter der Unruh mit 72 R gekennzeichnet ist.

Autavia case with screw back. The black dial with 3 white subdials. The 30-minute and 12-hour totalizers are positioned at 3 and 6 o'clock respectively. The permanent second subdial is positioned at 9 o'clock. The subdials are recessed with sloping sides and concentric circular ridges. A twin fully luminous marker is at 12 o'clock, with single luminous hour markers from 1 to 11, except at 3, 6, and 9 o'clock, which have luminous bars. The Autavia name is positioned over the Heuer logo. Pointed baton hands the same diameter as the hour markers have luminous inlays. The chronograph second hand is painted white, the small hands on the three subdials are black. "SWISS" and "T" for tritium reprinted on the hour subdial above 6 o'clock. The black minute bezel is bidirectional. A large crown has the Heuer name. The outer case back is signed "AUTAVIA HEUER" and "STAINLESS STEEL CHRONOGRAPH-WATERPROOF GUARANTEED 330FT UNDER WATER."
The interior of the case back and movement bridge are signed "HEUER-LEONIDAS SA." Powered by a Valjoux 72 movement, it is signed 72 R under the balance wheel.

Sekunden...
HEUER in allen Sportarten
seit Jahren dabei — mit Stoppuhren,
Chronographen und elektronischen
Zeitmessanlagen.
HEUER
1/10 SEC.
SWISS MADE
AUTAVIA
HEUER

AUTAVIA

ref. 2446 GMT

Autavia-Gehäuse mit verschraubtem Boden. Ein schwarzes Zifferblatt mit 3 weißen Hilfszifferblättern. Die 30-Minuten- und 12-Stunden-Totalisatoren befinden sich bei 3 bzw. 6 Uhr. Die permanente Sekunde als Hilfszifferblatt befindet sich bei 9 Uhr. Die Hilfszifferblätter sind vertieft, mit abgeschrägten Seiten und konzentrischen kreisförmigen Rillen.
Ein doppelter Leuchtindex befindet sich bei 12 Uhr, einzelne Leuchtindizes sind von 1 bis 11 angebracht, nur bei 3, 6 und 9 Uhr gibt es kleine Leuchtbalken. Die spitzen Stabzeiger haben Leuchteinlagen. Der 24-Stunden-GMT-Zeiger ist rot und hat eine mit Leuchtmasse besetzte Spitze. Der Sekundenzeiger des Chronographen ist weiß, die kleinen Zeiger auf den drei Hilfszifferblättern sind schwarz.
Auf dem Zifferblatt steht Autavia Heuer GMT (Greenwich Mean Time), über 6 Uhr ist „SWISS" aufgedruckt, und oben wurde ein kleines „T" hinzugefügt, das darauf hinweist, dass radioaktives Tritium als Leuchtmasse verwendet wurde.
Die rot-blaue Lünette, die Tag und Nacht anzeigt, ist in beide Richtungen drehbar und zeigt 24 Stunden an. Die Verwendung des roten GMT-Stundenzeigers ermöglicht das Ablesen einer zweiten Zeitzone. Die Krone ist mit einem Heuer-Logo versehen. Der äußere Gehäuseboden ist mit „AUTAVIA HEUER" und „STAINLESS STEEL CHRONOGRAPH-WATERPROOF GUARANTEED 330FT UNDER WATER" signiert.
Das Innere des Gehäusebodens und die Werkbrücke sind mit „HEUER-LEONIDAS SA" signiert. Angetrieben wird sie von einem Valjoux 724, das unter der Unruh mit 724 R gekennzeichnet ist.

Autavia case with sscrew back. The black dial has 3 white subdials. The 30-minute and 12-hour totalisers are positioned at 3 and 6 o'clock respectively. The permanent second subdial is positioned at 9 o'clock. The subdials are recessed with sloping sides and concentric circular ridges.
A twin fully luminous hour marker is at 12 o'clock, with single luminous hour markers from 1 to 11, except at 3, 6, and 9 o'clock, which have luminous bars. Pointed baton hands have luminous inserts. The 24-hour GMT hand is red with a luminous pointed tip. The second hand of the chronograph is white, the small hands on the three subdials are black.
The dial reads "AUTAVIA HEUER GMT" (Greenwich Mean Time), "SWISS" is printed above 6 o'clock, and a small T was added above, indicating that radioactive tritium was used as luminous material.
The red and blue day and night, bezel is bidirectional and indicates 24 hours. The use of the red GMT hour hand allows a second time zone to be followed. The crown has a Heuer logo. The outer case back is signed "AUTAVIA HEUER" and "STAINLESS STEEL CHRONOGRAPH-WATERPROOF GUARANTEED 330FT UNDER WATER."
The interior of the case back and movement bridge are signed "HEUER-LEONIDASSA." Powered by a Valjoux 724 movement, it is signed 724 R under the balance wheel.

AUTAVIA
HEUER
GMT
TAVIA 12
denzähler,
häuse,
30 Fuss),
der-
Ref. 2446 GMT
AUTAVIA GMT
1/5 Sekunden-Messer, 0–12 Stundenzähler,
schwarzes Zifferblatt, mit rotem
24 Stundenzeiger. Edelstahlgehäuse,
wasserdicht bi 100 m Tiefe
(330 Fuss), D hit 24 Stundenteilung.
Stahl- oder L d.
Er ist mit
zeiger verb
utomatisch
s vor jedem
reieck des
stellt wird.
West/Ost-F
fferenz in
rzeiger.
geht
6 Stunden au
Sie die Zeit i
urch den norm
n Paris (16.30
gleichzeitig a
d Sekundenzeig
leich.
ise von Paris n
hring sinngemä
gestellt.
man den
ronographen
ein
st im Fenster rech
nicht automatisc
sowohl beim Rich
tswechsel von Ha
den. Dies geschieh
ecknadel ganz leic
der sich seitlich a
efindet.
tag ist im Fenster li
tag-Scheibe spring
SUPER
AUTAVIA
HEUER
MADE IN SWITZERLAND

MARIO ANDRETTI
JACKY ICKX
AUTAVIA
HEUER
FAB.SUISSE
T
SWISS

AUTAVIA

ref. 3646 H „Mario Andretti“

Autavia-Gehäuse mit verschraubtem Gehäuseboden. Ein schwarzes Zifferblatt mit 2 weißen Hilfszifferblättern. Die permanente Sekunde und der 45-Minuten-Totalisator befinden sich bei 3 bzw. 9 Uhr. Die Hilfszifferblätter sind vertieft, mit abgeschrägten Seiten und konzentrischen kreisförmigen Rillen.
Ein großer Stahlindex befindet sich bei 12 Uhr, kleinere Stundenindizes aus Stahl sind von 1 bis 11 angebracht, außer bei 3 und 9 Uhr. Quadratische Leuchtpunkte befinden sich an der Außenseite der Indizes. Der Autavia-Schriftzug ist über dem Heuer-Logo positioniert.
Die spitzen Stabzeiger, die im Durchmesser mit den Indizes korrespondieren, sind mit Leuchteinlagen versehen. Der Chronographen-Sekundenzeiger ist weiß lackiert, die kleinen Zeiger auf den beiden Hilfszifferblättern sind schwarz.
„SWISS“ und „T“ für Tritium sind oberhalb von 6 Uhr aufgedruckt, darüber befindet sich der Schriftzug FAB.SUISSE. Dieser Schriftzug war verpflichtend bei jeder Uhr, die aus der Schweiz nach Frankreich exportiert wurde. Die schwarze Stunden-Lünette ist in beide Richtungen drehbar. Die Krone ist mit einem Heuer-Logo versehen. Der äußere Gehäuseboden ist mit „AUTAVIA HEUER“ und „STAINLESS STEEL CHRONOGRAPH-WATERPROOF GUARANTEED 330FT UNDER WATER“ signiert.
Das Innere des Gehäusebodens und die Werkbrücke sind mit „HEUER-LEONIDAS SA“ gekennzeichnet. Angetrieben wird sie von einem Valjoux 92 Uhrwerk, das unter der Unruh mit 92 R signiert ist.

Autavia case with screw back. The black dial has 2 white subdials. The permanent second and 45-minute totalizers are positioned at 3 and 9 o'clock respectively. The subdials are recessed with sloping sides and concentric circular ridges.
A large steel marker is at 12 o'clock, with smaller steel hour markers from 1 to 11, except at 3 and 9 o'clock. Luminous squares the outer side of the markers. The Autavia name is positioned over the Heuer logo.
Pointed baton hands the same diameter as the hour markers have luminous inlays. The chronograph second-hand is painted white; the small hands on the two subdials are black. "SWISS" and "T" for tritium is printed above 6 o'clock with a FAB.SUISSE (Swiss origin mark for import into France) mark on top. The black hour bezel is bidirectional. The crown has a Heuer logo. The outer case back is signed "AUTAVIA HEUER" and "STAINLESS STEEL CHRONOGRAPH-WATERPROOF GUARANTEED 330FT UNDER WATER."
The interior of the case back and movement bridge are signed "HEUER-LEONIDAS SA." Powered by a Valjoux 92 movement, it is signed 92 R under the balance wheel.

AUTAVIA

ref. 2446 H

Neues Gehäuse mit Druckboden. Modell mit schwarzem Zifferblatt und drei weißen Hilfszifferblättern. 30-Minuten- und 12-Stunden-Hilfszifferblätter bei 3 bzw. 6 Uhr, das permanente Sekunden-Hilfszifferblatt bei 9 Uhr. Die in eine Ebene versenkten Hilfszifferblätter haben konzentrische Rillen.
Die neu stilisierten Stundenzeiger sind aus Stahl mit schwarzen und reduzierten Leuchtmaterial-Einlagen. Die Stundenindizes haben gebürstete Oberflächen mit schwarzen Einschlüssen. Auf der äußeren 1/5-Sekunden-Skala steht bei jedem Index eine kleine rote Linie, die mit dem großen roten Sekundenzeiger übereinstimmt. Die kleinen Zeiger der drei Hilfszifferblätter sind schwarz. Der „AUTAVIA"-Schriftzug steht über dem Heuer-Logo, und bei den meisten Heuer-Uhren steht das Wort „SWISS" über 6 Uhr. Die schwarze Lünette ist in beide Richtungen drehbar, hat ein silbernes Dreieck bei 12 Uhr und zeigt Stundenmarkierungen. Die Krone ist größer und mit „HEUER" signiert. Größere Drücker mit Rillen wurden eingeführt. Der äußere Rückdeckel ist poliert und zeigt das Emblem der argentinischen Luftwaffe mit der militärischen Registriernummer. Das Uhrwerk ist ein Valjoux 72 und mit „72 R" unter der Unruhe signiert.

New snap-back case version. Model with black dial and three white subdials. The 30-minute and 12-hour totalizers are positioned at 3 and 6 o'clock respectively. The permanent second subdial is positioned at 9 o'clock. The subdials are recessed in the dial, with sloping sides and concentric circular ridges.
Newly designed steel hands have reduced black luminous inserts. Hour markers have brushed surfaces with black inserts. On the outer 1/5-second track at each marker is a fine red line matching the large red chronograph second hand; the small hands on the three subdials are black.
The "AUTAVIA" name is positioned over the Heuer logo, and most Heuer watches have "SWISS" above 6 o'clock.
The black bezel is bidirectional, with a silver triangle at 12 o'clock and hour markings. The crown is larger and signed "Heuer." Larger chronograph pushers with a fluted design were introduced. The outer back case is polished and displays the emblem and issue number of the Argentinian Air Force. The watch is powered by a Valjoux 72 movement and is signed "72 R" under the balance wheel.

AUTAVIA
HEUER
SWISS
T

AUTAVIA

ref. 2446 GMT

Gehäuse mit Druckboden. Modell mit schwarzem Zifferblatt und drei weißen Hilfszifferblättern. 30-Minuten- und 12-Stunden-Hilfszifferblätter bei 3 bzw. 6 Uhr, das permanente Sekunden-Hilfszifferblatt bei 9 Uhr. Die in einer Ebene versenkten Hilfszifferblätter haben konzentrische Rillen.

Polierte Stundenzeiger aus Stahl mit Leuchtmaterial-Einlagen. Der 24-Stunden-GMT-Zeiger ist rot mit Leuchtmaterial-Spitze. Der Chronographen-Sekundenzeiger ist weiß, die kleinen Zeiger der drei Hilfszifferblätter schwarz. Der „AUTAVIA"-Schriftzug steht über dem Heuer-Logo und dieses über „GMT" (Greenwich Mean Time). Die Stundenindizes sind Leuchtmaterial-Rechtecke. Die rote und blaue (Tag und Nacht) Lünette zeigt 24 Stunden an. Mit dem roten GMT-Zeiger kann eine zweite Zeitzone abgelesen werden. Die Krone ist mit „HEUER" signiert und die Drücker sind gerillt. Der Rückdeckel ist poliert. Die Firma Heuer begann in den späten 1960er Jahren, zwischen den Bandanstößen bei 12 Uhr die Modellnummer und die Seriennummer bei 6 Uhr einzustanzen.

Das Uhrwerk ist ein Valjoux 724, das vom Kaliber 72 abstammt und dem die GMT-Eigenschaft hinzugefügt wurde. Es ist unter der Unruh mit „R 724" signiert. Das Kompressor-Gehäuse ist federnd aufgehängt.

Snap-back case version. Model with black dial and three white subdials. The 30-minute and 12-hour totalizers are positioned at 3 and 6 o'clock respectively. The permanent second subdial is positioned at 9 o'clock. The subdials are recessed in the dial, with sloping sides and concentric circular ridges.

The polished steel hands have luminous inserts. The 24-hour GMT hand is red with luminous pointed tip. The second hand of the chronograph is white, the small hands on the three subdials are black.

The dial reads "AUTAVIA HEUER GMT" (Greenwich Mean Time). Hour markers are luminous squares. The red and blue day and night, bezel is bidirectional and indicates 24 hours. The use of the red GMT hour hand allows a second time zone to be followed. The large crown is signed "HEUER." The chronograph pushers are fluted. The back of the case is polished.

During the late 60s Heuer began to stamp the model reference between the lugs at 12 o'clock, and the serial number continues to be stamped at 6 o'clock.

The watch is powered by a Valjoux 724 movement, which is based on the Valjoux 72 movement with the addition of GMT feature and is signed "R 724" under the balance wheel. The compressor case has spring suspension.

AUTAVIA
HEUER
GMT
T
SWISS

AUTAVIA
HEUER
SWISS

AUTAVIA

ref. 73663 MH

Neu gestaltetes Autavia-Gehäuse mit verschraubtem Boden; Oberseite gebürstet und Seiten poliert. Ein schwarzes Zifferblatt mit 3 weißen Hilfszifferblättern. Die 30-Minuten- und 12-Stunden-Totalisatoren befinden sich bei 3 bzw. 6 Uhr. Die permanente Sekunde als Hilfszifferblatt ist bei 9 Uhr positioniert.
Die Hilfszifferblätter sind angeschrägt, mit Minuten- und Stundeneinteilung und konzentrischen kreisförmigen Rillen. Die Logos von Autavia und Heuer befinden sich unter dem 12-Uhr-Index.
Die Stabzeiger aus Stahl sind mit Leuchtmasse und schwarzen Einsätzen versehen. Die Stahlindizes haben weiße Einsätze und Leuchtbalken am äußeren Ende. Zwei parallele Indizes befinden sich bei 12 Uhr, bei 3, 6 und 9 Uhr wurde auf Indizes verzichtet.
Der rote Chronographen-Zeiger korrespondiert mit den roten Stundenmarkierungen auf der äußeren 1/5-Sekunden-Skala.
Die schwarze Lünette ist in beide Richtungen drehbar und verfügt über eine Rasterfunktion zur Anzeige von Minuten und Stunden (MH). Die Krone ist mit „HEUER" signiert und von integrierten geriffelten Chronographendrückern umgeben.
Das Uhrwerk ist ein Valjoux 7736 mit der Signatur „7736 R" unter der Unruh.

Newly designed Autavia screw-back case with brushed finish on top and polished sides. The dial is black with white 30-minute and 12-hour subdials at 3 and 6 o'clock respectively. A permanent second subdial is positioned at 9 o'clock.
The subdials are angled, with minute and hour divisions and concentric circular ridges. Autavia and Heuer logos are under the 12 o'clock index.
Steel baton hands have luminous and black inserts. Steel markers have white inserts and luminous bars at the outer end. Twin parallel markers are at 12 o'clock; markers have been omitted at 3, 6, and 9 o'clock.
The red chronograph hand matches the red hour markers on the outer 1/5-second track.
The black bezel is bidirectional with a ratchet mechanism indicating minutes and hours (MH). The crown is signed "HEUER" and is flanked by integrated fluted chronograph pushers.
The movement is a Valjoux 7736, and is signed "7736 R" under the balance wheel.

HEUER
1 Jahr DM 5,–
Taucher 100 m Tiefe
Flieger 10 000 m Höhe
AUTAVIA
HEUER
SWISS
HEUER

AUTAVIA

ref. 73463 „Orange Boy“ H.T.G Nr. 140

Autavia-Gehäuse mit verschraubtem Gehäuseboden; gebürstetes Finish auf der Oberseite und polierte Seiten. Das versilberte Zifferblatt hat zwei farblich abgesetzte bräunliche Hilfszifferblätter für die permanente Sekunde und den 30-Minuten-Zähler.
Die Hilfszifferblätter sind abgeschrägt und haben eine Sekunden- und Minutenteilung sowie konzentrische kreisförmige Rillen. Das „AUTAVIA“- und das Heuer-Logo befinden sich unter 12 Uhr und der Schriftzug „SWISS“ ist über dem silbern umrahmten Datumsfenster bei 6 Uhr angebracht.
Die orangefarbenen Zeiger haben Leuchteinlagen. Die Stahlindizes haben schwarze Einsätze und Leuchtbalken am äußeren Ende. Zwei parallele Indizes befinden sich bei 12 Uhr, bei 3, 6 und 9 Uhr wurde auf Indizes verzichtet. Der orangefarbene Chronographen-Zeiger endet genau bei den orangefarbenen Stundenmarkierungen auf der äußeren 1/5-Sekunden-Skala.
Die schwarze Minuten-Stunden-Lünette (MH) ist in beide Richtungen drehbar und verfügt über eine Rasterfunktion. Die Krone mit Heuer-Signatur und die geriffelten Chronographen-Drücker befinden sich auf der rechten Seite.
Der äußere Gehäuseboden ist mit der Gravur H.T.G. Nr. 140 versehen, was für die Heuer Time Gesellschaft steht, eine deutsche Heuer-Tochtergesellschaft. Nr. 140 ist die individuelle Nummer der Uhr, da diese für VIP-Kunden, Presse oder als kurzfristige Leihuhr während einer Reparatur oder Wartung verwendet wurden.
Die Uhr wird von einem Valjoux 7734-Handaufzugswerk angetrieben und ist auf der Brücke mit „HEUER LEONIDAS SA“ signiert.

Manual-wind Autavia screw-back case with brushed finish on top and polished sides. The silver-plated dial has contrasting brownish subdials for running second and 30-minute counters. Subdials are angled, with second and minute markers plus concentric circular ridges. "AUTAVIA" and the Heuer logo are under the 12 o'clock marker and "SWISS" is stamped above a silver-framed date window at 6 o'clock.
Orange hands have luminous inserts. Steel markers have black inserts and luminous bars at the outer end. Twin parallel markers are positioned at 12 o'clock; markers have been omitted at 3, 6 and 9 o'clock. The orange chronograph hand matches the orange hour markers on the outer 1/5 second track.
The black minute/hour bezel is bidirectional with a ratchet mechanism. The Heuer signed crown and fluted chronograph pushers are on the right side.
The outer case back is engraved with H.T.G. Nr.140, which stands for Heuer Time Gesellschaft, a German Heuer subsidiary. Nr.140 is the individual number of the watch as these were used for VIP clients, press, or short-term rental watch during service.
The watch is powered by a Valjoux 7734 manual movement and is signed "HEUER LEONIDAS SA" on the bridge.

AUTAVIA
HEUER
GMT
SWISS

AUTAVIA

ref. 741.603 GMT

Gehäuse mit verschraubtem Boden, Gehäuseoberseite mit Strichschliff und polierten Flanken. Das Zifferblatt ist schwarz und hat zwei weiße versenkte Hilfszifferblätter mit roten 5-Minuten-Markierungen auf dem Minuten-Hilfszifferblatt und nummerierten Stundenindizes von 1 bis 12 auf dem Stunden-Hilfszifferblatt. Ihre Flächen zeigen konzentrische Rillen.
„AUTAVIA", Heuer-Logo und „GMT" stehen unter 12 Uhr, „SWISS" über dem silbern gerahmten Datumsfenster bei 6 Uhr. Weiß bemalte Stahlzeiger mit roten Spitzen haben mit Leuchtmasse gefüllte Flächen. Der strahlend gelbe GMT-Zeiger dreht sich einmal in 24 Stunden. Die Indizes sind mit Leuchtmasse gefüllt. Ein doppelter Indexbalken ist bei 12 Uhr und bei 3, 6 und 9 Uhr fehlen die Indizes. Der rote Chronographen-Sekundenzeiger erreicht präzise die roten Stundenmarkierungen der 1/5-Sekunden-Skala. Die rote und blaue (Tag und Nacht) Lünette hat einen Leuchtpunkt mit darunterliegendem Dreieck und ist in beide Richtungen mit einer Rasterfunktion drehbar, sie zeigt 24 Stunden an. Die Krone ist mit „HEUER" signiert. Die Chronographen-Drücker sind gerillt und in das Gehäuse integriert. Die Uhr wird angetrieben von einem Valjoux 7741 Uhrwerk und ist mit „7741 R" unter der Unruhe signiert.

Screw-back case with brushed finish on top and polished sides. Black dial with two white recessed subdials with red 5-minute markers on the minute subdial and numbered hour markers from 1 to 12 on the hour register; they have concentric circular ridges. "AUTAVIA" and the Heuer logo are below 12 o'clock, with "GMT" below. "SWISS" is printed over the silver-framed date window at 6 o'clock.
White-painted steel hands with red tips have luminous inserts. The bright yellow GMT hand completes a single revolution every 24 hours. The markers are filled with luminous material. Twin parallel markers are positioned at 12 o'clock and markers have been omitted at 3, 6, and 9 o'clock. The red chronograph hand precisely reaches the red hour markers at the outer 1/5-second track.
The red and blue day and night bezel has a luminous dot and is bidirectional, with a click function indicating 24 hours.
The crown is signed "HEUER," the pushers are fluted and integrated into the case. The movement is a Valjoux 7741, and is signed "7741 R" under the balance wheel.

AUTAVIA

ref. 73663 MH

Größeres Gehäuse mit verschraubtem Boden, Gehäuseoberseite mit Strichschliff und polierten Flanken. Seltenes schwarzes Militär-Zifferblatt mit grünlichen Stundenindizes in arabischen Ziffern und drei schwarzen Hilfszifferblättern. 30-Minuten- und 12-Stunden-Hilfszifferblätter bei 3 bzw. 6 Uhr, das permanente Sekunden-Hilfszifferblatt bei 9 Uhr. Die Hilfszifferblätter sind auf gleicher Höhe mit dem Zifferblatt. „AUTAVIA" und Heuer-Logo stehen unter 12 Uhr. Die weißen Stahlzeiger haben große mit Leuchtmaterial gefüllte Flächen. Der weiße Chronographen-Sekundenzeiger hat an der Spitze ein mit Leuchtmasse gefülltes Dreieck. Am Rand des Zifferblattes ist eine Minuten- und Sekunden-Skala aufgedruckt.
Die schwarze Lünette ist in beide Richtungen durch eine Rasterfunktion drehbar und zeigt Minuten und Stunden. Die Krone ist mit „Heuer" signiert. Gerillte Chronographen-Drücker sind in das Gehäuse integriert.
Das Uhrwerk ist ein Valjoux Kaliber 7736 und auf der Brücke mit „7736 R" signiert.

Large newly designed screw-back case with brushed finish on top and polished sides. Rare black military dial with hour indices in greenish Arabic numerals and three black subdials.
The minute and hour totalizers are positioned at 3 and 6 o'clock respectively. The permanent second subdial is positioned at 9 o'clock. The subdials are flush with the dial. The "AUTAVIA" name is positioned over the Heuer logo.
White steel hands have large luminous inserts. The white chronograph second hand has a triangle with luminous insert at the tip.
At the edge of the dial there is a second and minute track.
The bezel is bidirectional with a click function indicating minutes and hours. The crown is signed "HEUER," the pushers are fluted and integrated into the case.
The movement is a Valjoux 7736, and is signed "7736 R" under the balance wheel.

AUTAVIA
HEUER
SWISS

Jo Siffert trägt eine HEUER Autavia Chronomatic, 1969.

Jo Siffert wears a HEUER Autavia Chronomatic, 1969.

Jo Siffert in der Box neben seinem March 701, beklebt mit HEUER-Logos, 1970.

Jo Siffert in the pits beside his March 701 adorned with HEUER logos in 1970.

TACHYMETER
CHRONOMATIC
HEUER
AUTAVIA
SWISS

AUTAVIA

ref. 1163 T „Jo Siffert“

Die Chronomatic ist das erste Automatik-Modell, oft von Sammlern als „Siffert“-Modell bezeichnet. Sie ist sehr selten, da sie nur wenige Monate in Produktion war. Das Zifferblatt ist weiß, mit schwarzen 30-Minuten- und 12-Stunden-Hilfszifferblättern. Das 12-Stunden-Hilfszifferblatt ist eine frühe Version und zeigt die Zahlen 3, 6, 9 und 12. Die Hilfszifferblätter sind bei den Minuten- und Stunden-Einteilungen abgeschrägt, die Planfläche zeigt konzentrische Rillen.
Der Schriftzug „CHRONOMATIC“ und das Heuer-Logo stehen unter 12 Uhr, „AUTAVIA“ über dem silbern gerahmten Datumsfenster, das Datum erscheint als schwarze Zahl bei 6 Uhr. Polierte Stahlzeiger haben mit Leuchtmasse gefüllte Flächen und zeigen gegen Stahlindizes mit Leuchtbalken am äußeren Ende. Zwei parallele Indizes sind bei 12 Uhr und keine bei 3, 6 und 9 Uhr. Der blaue Chronographen-Sekundenzeiger korrespondiert mit blauen Stundenmarkierungen an der 1/5-Sekunden-Skala. Die schwarze Tachymeter-Lünette ist in beide Richtungen mit einer Rasterfunktion drehbar und zeigt 63 bis 400 km/h. Die Krone ist links und mit „HEUER“ signiert. Gerillte Drücker sind rechts in das Gehäuse integriert. Der Boden ist verschraubt und hat einen Bürstenschliff.
Die Uhr wird angetrieben von einem frühen Kaliber 11 Automatik-Uhrwerk und ist auf der Brücke signiert.

The Chronomatic is the first automatic model, often referred to as the "Siffert" model by collectors. The Chronomatic is extremely rare as it was only produced for a few months. The dial is white with black 30-minute and 12-hour subdials. The hour markers are shown as 3, 6, 9, and 12, indicating an early dial variation. The subdials are angled at the minute and hour divisions; with concentric circular ridges on the faces.
"CHRONOMATIC" and the Heuer logo are below the 12 o'clock and "AUTAVIA" is printed over the silver-framed date window at 6 o'clock. Polished steel hands incorporate cast luminous material; steel indices have luminous bars on their outer ends. Twin parallel markers are positioned at 12 o'clock and markers have been omitted at 3, 6, and 9 o'clock. The blue chronograph hand matches the blue hour markers of the 1/5-second track.
The black tachymeter bezel is bidirectional with a click function indicating 63 to 400 km/h (39-249 mph). The crown is on the left-hand side and is signed "HEUER." Fluted chronograph pushers are integrated into the case on the right-hand side. The case back is screw-back with brushed finish.
The watch is powered by an early Calibre 11 automatic movement and is signed on the bridge.

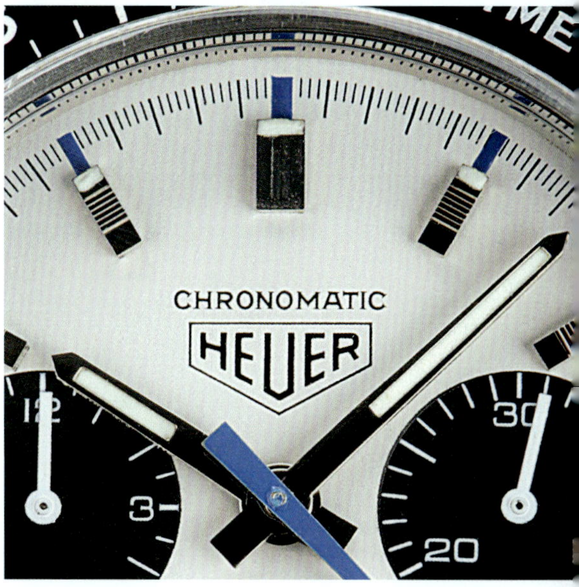

Weltklassefahrer
Jo Siffert
trägt den HEUER
Chronographen
mit Automatik.
TACHYMETER
AUTAVIA
HEUER
AUTOMATIC CHRONOGRAPH
Heuer Autavia Ref. 1163T
Chronograph, stoppt 1/5 Sekunde,
Monte Carlo Ref. 542.701
Standard-Zifferblatt, 1/5 Sekunden-Messer,
HEUER, der führende Hersteller von Chronographen, Stoppuhren und Rally-Uhren. Bitte fordern Sie Informationsmaterial über das gesamte HEUER-Programm an. Rennaufkleber senden wir Ihnen gerne gratis.
HEUER
CHRIST
IWC Uhren
Frankfurt am

AUTAVIA

ref. 1163 T „Jo Siffert“

Erstes Modell nach der Chronomatic, bei dem der Autavia-Schriftzug über dem Heuer-Logo neu positioniert wurde, von Sammlern oft als „Jo Siffert“-Modell bezeichnet. Das Zifferblatt ist cremefarben mit schwarzen 30-Minuten- und 12-Stunden-Hilfszifferblättern.
Das 12-Stunden-Hilfszifferblatt mit den Aufdrucken 3, 6, 9 und 12 weist auf die frühe Zifferblattvariante hin, während die späteren Modelle die Stunden von 1 bis 12 anzeigen.
Die Hilfszifferblätter sind angeschrägt, mit Minuten- und Stunden-Einteilung und konzentrischen kreisförmigen Rillen.
Unter 12 Uhr steht Autavia (seltener flacher V-Aufdruck bei diesem Übergangsmodell) und das Heuer-Logo, sowie „AUTOMATIC CHRONOGRAPH“ über einem Datumsfenster mit silberfarbenem Rahmen bei 6 Uhr. Polierte Stahlzeiger mit Leuchteinlagen zeigen auf Zifferblattindizes aus Stahl mit Leuchtbalken an den äußeren Enden. Zwei parallele Indizes befinden sich bei 12 Uhr, während bei 3, 6 und 9 Uhr auf Indizes verzichtet wurde. Der blaue Chronographen-Zeiger korrespondiert mit breiteren (bei diesem frühen Exemplar) blauen Stundenmarkierungen auf der äußeren 1/5-Sekunden-Skala.
Die schwarze Tachymeter-Lünette ist in beide Richtungen drehbar, mit einer Rasterfunktion, die 63 bis 400 km/h anzeigt. Die Krone befindet sich auf der linken Seite und ist mit „HEUER“ signiert. Die geriffelten Drücker auf der rechten Seite sind (bei diesem frühen Gehäuse) kaum integriert. Der Gehäuseboden ist verschraubt und gebürstet.
Die Uhr wird von einem frühen Automatikwerk Kaliber 11 angetrieben und ist auf der Brücke mit „CAL 11“ signiert.

The first model after the Chronomatic to show Autavia repositioned above the Heuer logo, often referred to as the "Jo Siffert" model by collectors. Cream dial with black 30-minute and 12-hour subdials.
The hour subdial printed 3, 6, 9, and 12 indicates an early dial variation; later ones have hours marked from 1 to 12.
The subdials are angled with minute and hour divisions and concentric circular ridges.
The Autavia name (rare flat V print on this transitional piece) and Heuer logo are underneath the 12 o'clock and "AUTOMATIC CHRONOGRAPH" is printed over a silver-framed date window at 6 o'clock. Polished steel hands with luminous inlays track steel markers on the dial with luminous bars on the outer ends. Twin parallel markers are positioned at 12 o'clock; markers have been omitted at 3, 6, and 9 o'clock. The blue chronograph hand matches the wider (on this early piece) blue hour markers on the outer 1/5-second track.
The black tachymeter bezel is bidirectional with a click function indicating 63 to 400 km/h (39-249 mph). The crown is on the left-hand side and is signed "HEUER". Fluted pushers are barely (on this early case) integrated on the right-hand side. The case back is screw-back with brushed finish.
The watch is powered by an early Calibre 11 automatic movement and is, signed "CAL 11" on the bridge.

AUTAVIA

ref. 1163 MH

Automatik-Gehäuse mit verschraubtem Boden, Gehäuseoberseite mit Strichschliff und polierten Flanken. Das Zifferblatt ist schwarz mit weißen 30-Minuten- und 12-Stunden-Hilfszifferblättern. Das 12-Stunden-Hilfszifferblatt, zeigt die Zahlen 3, 6, 9 und 12. Die Hilfszifferblätter sind bei den Minuten- und Stunden-Einteilungen abgeschrägt, die Planfläche zeigt konzentrische Rillen.

„AUTAVIA“ und Heuer-Logo mit neuer Schrifttype stehen unter 12 Uhr, „AUTOMATIC CHRONOGRAPH“ über dem silbern gerahmten Datumsfenster bei 6 Uhr.

Polierte Stahlzeiger haben mit Leuchtmasse gefüllte Flächen und zeigen gegen die Mitte hin schwarze Einlagen. Die Stahlindizes auf dem schwarzen Zifferblatt wurden neu gestaltet und erhielten weiße Einschlüsse sowie einen Leuchtbalken am äußeren Ende. Der rote Chronographen-Sekundenzeiger korrespondiert mit roten Stundenmarkierungen an der 1/5-Sekunden-Skala. Die schwarze Lünette mit Rasterfunktion ist in beide Richtungen drehbar und zeigt Minuten und Stunden. Die Krone ist links und mit „HEUER“ signiert. Gerillte Chronographen-Drücker sind rechts in das Gehäuse integriert. Die Uhr wird angetrieben von einem Kaliber 11 Automatik-Uhrwerk und ist auf der Brücke signiert.

Automatic screw-back case with brushed finish on top and polished sides. The dial is black with white 30-minute and 12-hour subdials. The hour markers are printed 3, 6, 9, and 12, indicating an early dial variation.

The subdials are angled at the minute and hour divisions; the face is flat with concentric circular ridges.

The Autavia and Heuer logos in new typeface are below 12 o'clock and "AUTOMATIC CHRONOGRAPH" is printed above the silver-framed date window at 6 o'clock.

Polished steel hands incorporate cast luminous material and small black inserts towards the center. The steel markers on the black dial were redesigned with white inserts and luminous bars at the outer end. The red chronograph hand matches the red hour markers of the outer 1/5-second track. The black bezel is bidirectional with a click function indicating minutes and hours. The crown is on the left-hand side and is signed "HEUER." Fluted chronograph pushers are integrated into the case on the right-hand side.

The watch is powered by a Calibre 11 automatic movement and is signed on the bridge.

AUTAVIA
HEUER
AUTOMATIC
CHRONOGRAPH
SWISS

Get a taste of what it's all about.
Get the full taste of Viceroy.
Castrol
VICEROY
Light OLYMPIA BEER
AUTAVIA
HEUER
TACHYMETER
This $200 HEUER chronograph—yours for only $88.
Now wear the Heuer self-winding chronograph of the Parnelli Jones—Viceroy Racing Team for $88. This 17-jewel watch/stopwatch features: stainless-steel water-resistant case, tachymeter bezel, self-winding movement, date window, recording register from 0-12 hours with a 1/5 second sweep hand.
Heuer is the official timing instrument at major international races including the U.S. Grand Prix and Sebring 12-hour Endurance Race. To get this $200 watch with gift case, just send 1 Viceroy carton end flap and $88 to:
Parnelli Jones
Viceroy-Heuer Chronograph Offer
Box 3000 B
Louisville, Kentucky 40201
☐ Check ☐ Money Order ☐ Master Charge
☐ BankAmericard
ACCT. NUMBER
NAME
I AM OVER 21 YEARS OF AGE
ADDRESS
CITY STATE ZIP
L
Please allow 6 weeks for delivery. Void where taxed, prohibited or restricted. Offer expires Dec. 31, 1972.

Gulf Porsche 917K, gefahren von Derek Bell und Jo Siffert, 1971.

Gulf Porsche 917K driven by Derek Bell and Jo Siffert in 1971.

Links: HEUER und das Parnelli Jones-Viceroy-Racing Team werben für die Autavia 1163 V, genannt „Viceroy"

Left: HEUER and Parnelli Jones-Viceroy Racing Team promote the Autavia 1163 V, „Viceroy."

TACHYMETER
AUTAVIA
HEUER
AUTOMATIC
CHRONOGRAPH
SWISS
18

AUTAVIA

ref. 1163 T „Orange Boy“

Autavia Automatik-Gehäuse mit verschraubter Rückseite, gebürsteter Oberseite und polierten Seiten. Das Zifferblatt ist von weiß zu eierschalenfarben gealtert, die zwei Hilfszifferblätter (30-Minuten und 12-Stunden) sind von schwarz auf braun verblichen.
Das 12-Stunden-Hilfszifferblatt ist eine spätere Version mit Nummerierung von 1 bis 12.
Das 30-Minuten-Hilfszifferblatt ist mit 3 orangen Balken versehen, jeder Balken markiert 5 Minuten.
Die Hilfszifferblätter sind angeschrägt, mit Minuten- und Stunden-Einteilung und konzentrischen kreisförmigen Rillen. Das Autavia- und das Heuer-Logo befinden sich unter dem 12-Uhr-Index, der Schriftzug „AUTOMATIC CHRONOGRAPH“ ist über dem Datumsfenster mit silberfarbenem Rahmen bei 6 Uhr aufgedruckt.
Die Zeiger aus gebürstetem Stahl sind mit Leuchtmasse und orangefarbenen Einsätzen in der Mitte sowie orangefarbenen Spitzen versehen. Die Stahlindizes haben schwarze Einsätze und Leuchtbalken am äußeren Ende. Zwei parallele Indizes befinden sich bei 12 Uhr, während die Indizes bei 3, 6 und 9 Uhr weggelassen wurden. Der orangefarbene Chronographen-Zeiger korrespondiert mit den orangefarbenen verblassten Stundenmarkierungen auf der äußeren 1/5-Sekunden-Skala.
Die schwarze Tachymeter-Lünette ist in beide Richtungen drehbar und verfügt über eine Rasterfunktion. Die Krone befindet sich auf der linken Seite und ist mit „HEUER“ signiert. Die integrierten geriffelten Chronographen-Drücker befinden sich auf der rechten Seite. Das Gehäuse ist verschraubt und gebürstet.
Die Uhr wird von einem Automatik-Werk Kaliber 12 angetrieben und ist auf der Brücke mit „CAL 12“ signiert.

Automatic Autavia screw-back case with brushed finish on top and polished sides. The dial has aged from white to eggshell, the 30-minute and 12-hour subdials from black to brown.
The hour subdial is a later version numbered from 1 to 12; the 30-minute counter has three orange markers indicating 5 minutes each.
Subdials are angled, with minute and hour divisions and concentric circular ridges. The Autavia and Heuer logos are under the 12 o'clock marker and "AUTOMATIC CHRONOGRAPH" is printed over the silver-framed date window at 6 o'clock.
Brushed steel hands have luminous and orange inserts towards the center and orange tips. Steel indices have black inserts and luminous bars at the outer end. Twin parallel markers are positioned at 12 o'clock and indices have been omitted at 3, 6, and 9 o'clock. The orange chronograph hand matches the orange faded hour markers on the outer 1/5-second track.
The black tachymeter bezel is bidirectional with a click function. The crown is on the left side and signed "HEUER." Integrated fluted chronograph pushers are on the right.
The watch is powered by a Calibre 12 automatic movement and is signed on the bridge.

AUTAVIA

ref. 1163 GMT „Red Needles“

Autavia Automatik-Gehäuse, verschraubter Boden, gebürstetes Finish auf der Oberseite und größere, polierte Flanken. Braunes Farbwechselzifferblatt mit zwei weißen versenkten Hilfszifferblättern, nummerierte Stundenindizes von 1 bis 12 und 10, 20, 30 auf dem Minuten-Hilfszifferblatt. Die Zeiger der Hilfszifferblätter sind bei dieser nur kurzzeitig erhältlichen Version rot.
Autavia- und Heuer-Logo stehen über dem GMT-Schriftzug unterhalb von 12 Uhr. Der Schriftzug „AUTOMATIC CHRONOGRAPH“ ist über dem Datumsfenster mit silberfarbenem Rahmen bei 6 Uhr aufgedruckt.
Die Zeiger aus gebürstetem Stahl sind mit Leuchtmasse und roten Einsätzen in der Mitte sowie roten Spitzen versehen. Der leuchtend gelbe GMT-Zeiger, dessen Spitzen mit Leuchtmasse versehen sind, vollzieht eine Umdrehung in 24 Stunden. Die Indizes sind mit Leuchtmasse ausgefüllt. Zwei parallele Indizes befinden sich bei 12 Uhr, bei 3, 6 und 9 Uhr wurde auf Indizes verzichtet. Der rote Chronographenzeiger erreicht präzise die roten Stundenmarkierungen auf der äußeren 1/5-Sekunden-Skala.
Die verblasste rot-blaue Tag- und Nacht-Lünette hat einen Leuchtpunkt, ist in beide Richtungen drehbar und zeigt 24 Stunden an. Die Krone befindet sich auf der linken Seite und ist mit „HEUER“ signiert, die Drücker befinden sich auf der rechten Seite.
Die Uhr wird von einem Automatikwerk Kaliber 14 angetrieben und ist auf der Brücke mit „CAL 14“ signiert.

Automatic Autavia case design, screw back, brushed finish on top and larger polished sides. Brown color-change dial with two white recessed subdials, numbered hour markers from 1 to 12, and 10, 20, 30 on the minute register. The subdial hands are red on this short-lived version.
The Autavia name and Heuer logo are underneath 12 o'clock and GMT is added below; "AUTOMATIC CHRONOGRAPH" is printed over the silver-framed date window at 6 o'clock.
Brushed steel hands have luminous and red inserts towards the center and red tips. The bright yellow LUMINOUS-TIPPED GMT hand completes a single revolution every 24 hours. The markers are of luminous material. Twin parallel markers are positioned at 12 o'clock and markers have been omitted at 3, 6, and 9 o'clock. The red chronograph hand precisely reaches the red hour markers at the outer 1/5-second track.
The faded red and blue (day and night) bezel has a luminous dot and is bidirectional, indicating 24 hours. The crown is on the left and signed "HEUER" with pushers on the right.
The watch is powered by a Calibre 14 automatic movement and is signed "CAL 14" on the bridge.

AUTAVIA
HEUER
GMT
AUTOMATIC
CHRONOGRAPH
SWISS
UNITED STATES OF AMERICA
UNITED STATES OF AMERICA

Zeit-Roboter
TACHYMETER
AUTAVIA
HEUER
AUTOMATIC
CHRONOGRAPH
SWISS

AUTAVIA

ref. 1163 T „Jo Siffert“

Verschraubtes Autavia Automatik-Gehäuse mit gebürsteter Oberseite und polierten Seiten. Das Zifferblatt ist cremefarben mit schwarzen 30-Minuten- und 12-Stunden-Hilfszifferblättern.
Das Stundenzifferblatt ist eine spätere Version mit einer Nummerierung von 1 bis 12.
Die Hilfszifferblätter sind angeschrägt, mit Minuten- und Stundeneinteilung und konzentrischen kreisförmigen Rillen. Das Autavia- und das Heuer-Logo befinden sich unter dem 12-Uhr-Index, und der Schriftzug „AUTOMATIC CHRONOGRAPH“ ist über dem silbern umrahmten Datumsfenster bei 6 Uhr aufgedruckt.
Die Zeiger aus gebürstetem Stahl sind mit Leuchtmasse und blauen Einsätzen in der Mitte sowie blauen Spitzen versehen. Die Stahlindizes haben schwarze Einsätze und Leuchtbalken am äußeren Ende. Zwei parallele Indizes befinden sich bei 12 Uhr, während die Indizes bei 3, 6 und 9 Uhr weggelassen wurden. Der blaue Chronographenzeiger korrespondiert mit den blauen Stundenmarkierungen auf der äußeren 1/5-Sekunden-Skala.
Die schwarze Tachymeter-Lünette (T) ist in beide Richtungen drehbar und verfügt über eine Rasterfunktion. Die Krone befindet sich auf der linken Seite und ist mit „HEUER“ signiert. Die integrierten geriffelten Chronographen-Drücker befinden sich auf der rechten Seite. Das Gehäuse ist mit einem Schraubverschluss versehen und gebürstet.
Die Uhr wird von einem Automatikwerk Kaliber 12 angetrieben und ist auf der Brücke „CAL 12“ signiert.

Automatic Autavia screw-back case with brushed finish on top and polished sides. The cream dial has black 30-minute and 12-hour subdials.
The hour subdial is a later version numbered from 1 to 12.
Subdials are angled, with minute and hour divisions and concentric circular ridges. The Autavita and Heuer logos are under the 12 o'clock marker and "AUTOMATIC CHRONOGRAPH" is printed over the silver-framed date window at 6 o'clock.
Brushed steel hands have luminous and blue inserts towards the center and blue tips. Steel markers have black inserts and luminous bars at the outer end. Twin parallel markers are positioned at 12 o'clock and markers have been omitted at 3, 6, and 9 o'clock. The blue chronograph hand matches blue hour markers on the outer 1/5-second track.
The black tachymeter (T) bezel is bidirectional with a click function. The crown is on the left and signed "HEUER." Integrated fluted chronograph pushers are on the right.
The watch is powered by a Calibre 12 automatic movement and is signed "CAL 12" on the bridge.

AUTAVIA
HEUER
AUTOMATIC
CHRONOGRAPH
SWISS
18
AUTAVIA
HEUER

AUTAVIA

ref. 1163 MH H.T.G. Nr. 122

Autavia Automatik-Gehäuse, verschraubter Boden, gebürstetes Finish auf der Oberseite und polierte Seiten. Schwarzes Zifferblatt mit weißen 30-Minuten- und 12-Stunden-Hilfszifferblättern.
Die Hilfszifferblätter sind angeschrägt, mit Minuten- und Stunden-Einteilung und konzentrischen kreisförmigen Rillen. Das Autavia- und das Heuer-Logo befinden sich unter 12 Uhr, und der Schriftzug „AUTOMATIC CHRONOGRAPH“ ist über dem Datumsfenster mit silberfarbenem Rahmen bei 6 Uhr aufgedruckt.
Die Zeiger aus gebürstetem Stahl haben in der Mitte große Einsätze mit Leuchtmasse und roten Spitzen. Die Stahlindizes haben weiße Einsätze und Leuchtbalken am äußeren Ende. Zwei parallele Indizes befinden sich bei 12 Uhr, die Indizes bei 3, 6 und 9 Uhr wurden weggelassen. Der rote Chronographenzeiger korrespondiert mit den roten Stundenmarkierungen auf der äußeren 1/5-Sekunden-Skala.
Die Minuten- und Stunden-Lünette (MH) ist in beide Richtungen drehbar und verfügt über eine Rasterfunktion. Die Krone befindet sich auf der linken Seite und ist mit „HEUER“ signiert. Die integrierten geriffelten Chronographen-Drücker befinden sich auf der rechten Seite. Der äußere Gehäuseboden ist mit der Gravur H.T.G. Nr 122 versehen, die für die Heuer Time Gesellschaft steht, eine deutsche Heuer-Tochtergesellschaft. Nr 122 ist die individuelle Nummer der Uhr, da diese für VIP-Kunden, die Presse oder als Leihuhr während einer Reparatur verwendet wurde.
Die Uhr wird von einem Heuer-Automatikwerk angetrieben, das auf der Brücke signiert ist.

Automatic Autavia screw-back case with brushed finish on top and polished sides. A black dial has white 30-minute and 12-hour subdials.
Subdials are angled, with minute and hour divisions and concentric circular ridges. The Autavia and Heuer logos are under the 12 o'clock index and "AUTOMATIC CHRONOGRAPH" is printed over the silver-framed date window at 6 o'clock.
Brushed steel hands have large luminous and red inserts towards the centre and red tips. Steel markers have white inserts and luminous bars at the outer end. Twin parallel markers are positioned at 12 o'clock; markers have been omitted at 3, 6, and 9 o'clock. The red chronograph hand matches the red hour markers on the outer 1/5-second track.
The minute and hour (MH) bezel is bidirectional with a click function. The crown is on the left side and signed "HEUER." Integrated fluted chronograph pushers are on the right. The outer case back is engraved H.T.G. Nr 122, which stands for Heuer Time Gesellschaft, a German Heuer subsidiary. Nr 122 is the individual number of the watch, as these were used for VIP clients, press, or short-term rental during service.
The watch is powered by a Heuer automatic movement, signed on the bridge.

AUTAVIA
HEUER
GMT
AUTOMATIC
CHRONOGRAPH
SWISS

AUTAVIA

ref. 1163 GMT

Autavia Automatik-Gehäuse, verschraubter Boden, gebürstetes Finish auf der Oberseite und polierte Seiten. Schwarzes Zifferblatt mit zwei weißen vertieften Hilfszifferblättern mit konzentrischen kreisförmigen Rillen, roten 5-Minuten-Markierungen auf dem Minuten-Hilfszifferblatt und Indizes von 1 bis 12 auf dem Stunden-Hilfszifferblatt.
Das Autavia- und das Heuer-Logo befinden sich unterhalb von 12 Uhr und über dem GMT-Logo. „AUTOMATIC CHRONOGRAPH“ ist über dem Datumsfenster mit silberfarbenem Rahmen bei 6 Uhr aufgedruckt.
Die rot lackierten Stahlzeiger haben große Leuchteinlagen. Der leuchtend gelbe GMT-Zeiger mit quadratischer Spitze vollführt in 24 Stunden eine volle Umdrehung. Die Indizes sind mit Leuchtmasse gefüllt. Zwei parallele Indizes befinden sich bei 12 Uhr, bei 3, 6 und 9 Uhr wurde auf Indizes verzichtet. Der rote Chronographen-Zeiger erreicht präzise die roten Stundenmarkierungen auf der äußeren 1/5-Sekunden-Skala.
Die rot-blaue 24-Stunden-Lünette (Tag und Nacht) hat einen Leuchtpunkt und ist in beide Richtungen drehbar. Die Krone befindet sich auf der linken Seite und ist mit „HEUER“ signiert. Die Drücker sind ins Gehäuse integriert.
Die Uhr wird von einem Automatikwerk Kaliber 14 angetrieben und ist auf der Brücke „CAL. 14“ signiert.

Automatic Autavia case, screw back, brushed finish on top, and polished sides. Black dial with two white recessed subdials, red 5-minute marks on the minute subdial, hour indices from 1 to 12 on the hour register, and concentric circular ridges.
The Autavia and Heuer logos are underneath 12 o'clock and GMT is added beneath; "AUTOMATIC CHRONOGRAPH" is printed over the silver-framed date window at 6 o'clock.
Red-painted steel hands have large luminous inserts. The bright yellow GMT hand with square tip completes a single revolution every 24 hours. The markers are filled with luminous material. Twin parallel markers are positioned at 12 o'clock and markers have been omitted at 3, 6, and 9 o'clock. The red chronograph hand precisely reaches the red hour markers at the outer 1/5-second track.
The red and blue (day and night) bezel has a luminous dot and is bidirectional, with a click function denoting 24 hours. The crown is on the left side and signed "HEUER." The pushers are integrated into the case.
The watch is powered by a Calibre 14 automatic movement and is signed "CAL 14" on the bridge.

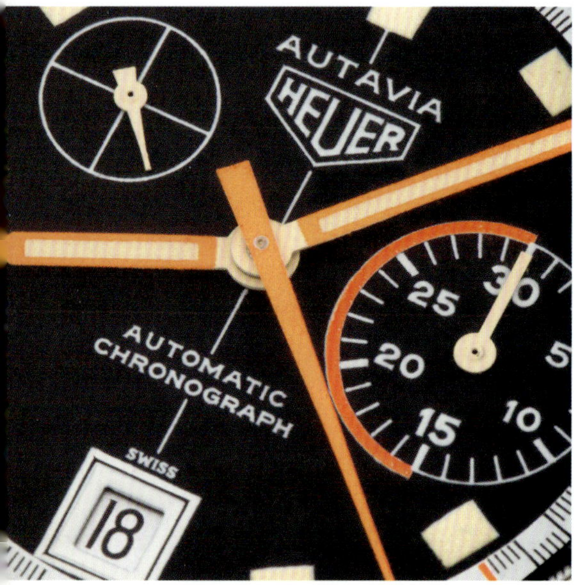

AUTAVIA

ref. 1163 V „Exotic“

Autavia Automatik-Gehäuse, verschraubter Boden, gebürstetes Finish auf der Oberseite und polierte Seiten. Das schwarze Zifferblatt, mit beigefarbener kleinen Sekunde bei 10 und einem Kreis im Stile einer Zielscheibe, ist eine Augenweide. Das 30-Minuten-Hilfszifferblatt bei 3 Uhr ist mit Ziffern im 5-Minuten-Takt und einem ungewöhnlichen orangefarbenen Halbkreis zwischen 15 und 30 versehen.
Autavia- und Heuer-Logo befinden sich unter dem 12-Uhr-Index, der Schriftzug „AUTOMATIC CHRONOGRAPH“ ist über dem silbern gerahmten Datumsfenster bei 6 Uhr aufgedruckt. Eine weiße Linie zieht sich von 12 nach 6 Uhr.
Die orangenen Zeiger haben große Leuchteinlagen. Die rechteckigen Leuchtindizes sind zur besseren Ablesbarkeit bei schlechten Lichtverhältnissen vergrößert. Ein großer quaderförmiger Index befindet sich bei 12 Uhr. Die Indizes wurden weggelassen bei 3, 6 und 9 Uhr. Der orangefarbene Chronographen-Zeiger korrespondiert mit den orangefarbenen Stundenmarkierungen und einer äußeren 1/5-Sekunden-Skala im Stil einer Ziellinie.
Die Tachymeter-Lünette ist in beide Richtungen drehbar und verfügt über eine Rasterfunktion. Die Krone befindet sich auf der linken Seite und ist mit „HEUER“ signiert. Die integrierten geriffelten Chronographen-Drücker sind auf der rechten Seite positioniert. Der innere Gehäuseboden ist mit „HEUER LEONIDAS“ signiert.
Die Uhr wird von einem automatischen Heuer-Kaliber 15 angetrieben, auf der Brücke signiert mit „HEUER CAL. 15“.

Automatic Autavia screw-back case with brushed finish on top and polished sides. The all-black dial is stunning, featuring a small beige running second hand at 10 o'clock in a target-style circle. A 30-minute sub dial at 3 o'clock has 5-minute numbers and an unusual orange printed semicircle between 15 to 30.
The Autavia and Heuer logos are under the 12 o'clock marker; "AUTOMATIC CHRONOGRAPH" is printed over the silver-framed date window at 6 o'clock with a vertical a white line between 12 and 6 o'clock.
Orange-painted hands have large luminous inserts. Luminous rectangular indices are enlarged for better visibility in poor lighting conditions. A large cuboid marker is positioned at 12 o'clock. Hour markers have been omitted at 3, 6, and 9 o'clock. The orange chronograph hand matches the orange hour markers and the outer 1/5-second track is designed like a finishing line.
The tachymeter bezel is bidirectional with a click function. The crown is on the left side and signed "HEUER." Integrated fluted chronograph pushers are on the right side. The inner case back is signed "HEUER LEONIDAS."
The watch is powered by a Heuer Calibre 15 automatic movement, signed "HEUER CAL. 15" on the bridge.

HEUER
AUTAVIA
HEUER
AUTOMATIC
CHRONOGRAPH
SWISS
18
TACHYMETER
HEUER

MODELL AUTAVIA DEKOMPRESSION
MODÈLE AUTAVIA DÉCOMPRESSION
AUTAVIA DECOMPRESSION MODEL
METERS
AUTAVIA
HEUER
AUTOMATIC
CHRONOGRAPH
SWISS
METERS

AUTAVIA

ref. 11630 P „Decompression Model“

Automatik-Autavia mit in der Tiefe vergrößertem Gehäuse und verschraubtem Stahlboden, oben gebürstet, und breiteren, polierten Flanken. Braunes Zifferblatt mit zwei weißen vertieften Hilfszifferblättern, die konzentrische kreisförmige Rillen aufweisen. Orangefarbene 5-Minuten-Markierungen auf dem Minuten-Hilfszifferblatt, Indizes von 1, 2, 3 bis 12 auf dem Stunden-Hilfszifferblatt.
Der Modellname Autavia ist im Vergleich zum Heuer-Logo unterhalb von 12 Uhr vergrößert dargestellt. Der Schriftzug „AUTOMATIC CHRONOGRAPH“ ist über dem Datumsfenster bei 6 Uhr aufgedruckt.
Die orangefarbenen Zeiger haben große Leuchteinlagen. Die Indizes sind mit Leuchtmasse gefüllt. Zwei parallele Indizes befinden sich bei 12 Uhr, bei 3, 6 und 9 Uhr wurde auf Indizes verzichtet. Der orangefarbene Chronographen-Zeiger erreicht präzise die orangefarbenen Stundenmarkierungen auf der äußeren 1/5-Sekunden-Skala.
Ein mattschwarzer Spannring ist zwischen dem Zifferblatt und dem Mineralglas befestigt.
Die schwarze Dekompressions-Lünette mit Leuchtpunkt ist in beide Richtungen drehbar und verfügt über eine Skala mit Meterangabe, um die Zeit für die Dekompression in einer bestimmten Tiefe zu berechnen. Die Krone befindet sich auf der linken Seite und ist mit „HEUER“ signiert. Die Drücker sind auf der Vorderseite integriert, aber auf der Rückseite vollständig sichtbar.
Der äußere Gehäuseboden ist mit der Gravur H.T.G. NR. 101 versehen, was für die Heuer Time Gesellschaft steht, eine deutsche Heuer-Tochtergesellschaft.
Die Uhr wird von einem Automatikwerk Kaliber 12 angetrieben und ist auf der Brücke mit „HEUER CAL.12“ signiert.

Automatic Autavia deep case design with steel screw back, brushed finish on top, and wider polished sides. Brown dial with two white recessed subdails with concentric circular ridges, orange 5-minute marks on minute subdial, indices from 1,2,3 to 12 on the hour register.
The Autavia name is enlarged compared to the Heuer logo underneath 12 o'clock and "AUTOMATIC CHRONOGRAPH" is printed over the date window at 6 o'clock.
Orange-painted hands have large luminous inserts. Markers are filled with luminous material. Twin parallel markers are positioned at 12 o'clock; markers have been omitted at 3, 6, and 9 o'clock. The orange chronograph hand precisely reaches the orange hour markers at the outer 1/5-second track.
A matte black tension ring is fixed between the dial and the mineral glass.
The black decompression bezel has a luminous dot and is bidirectional with a click function indicating meters to calculate the amount of time for decompression at a particular depth. The crown is on the left side and signed "HEUER." The pushers appear integrated from the front but are fully visible from the back.
The outer case back is engraved H.T.G. Nr. 101, which stands for Heuer Time Gesellschaft, a German Heuer subsidiary.
The watch is powered by a Calibre 12 automatic movement and is signed on the bridge.

AUTAVIA

ref. 11630 GMT

Automatik mit neuem vertieftem Gehäuse-Design, verschraubtem Boden, Gehäuseoberseite mit Strichschliff und vergrößerten polierten Flanken. Das Zifferblatt ist schwarz und hat zwei graue versenkte Hilfszifferblätter mit orangefarbigen 5-Minuten-Markierungen auf dem Minuten-Hilfszifferblatt und nummerierte Stundenindizes von 1 bis 12 am 12-Stunden-Hilfszifferblatt. Ihre Flächen zeigen konzentrische Rillen.
„AUTAVIA", Heuer-Logo und „GMT" stehen unter 12 Uhr. „AUTOMATIC CHRONOGRAPH" steht über dem silbern gerahmten Datumsfenster bei 6 Uhr. Weiß bemalte Stahlzeiger haben mit Leuchtmasse gefüllte Flächen. Der strahlend gelbe GMT-Zeiger dreht sich einmal in 24 Stunden. Die Indizes sind mit Leuchtmasse gefüllt. Ein doppelter Index bei 12 Uhr und keine bei 3, 6 und 9 Uhr. Der orange Chronographen-Sekundenzeiger erreicht präzise die orangen Stundenmarkierungen der 1/5-Sekunden-Skala. Ein großer schwarzer Spannring wird zwischen Zifferblatt und Glas fixiert. Die rote und blaue (Tag und Nacht) Lünette hat einen Leuchtpunkt, eine Rasterfunktion, ist in beide Richtungen drehbar und zeigt 24 Stunden an. Die Krone ist links und mit „HEUER" signiert. Die Chronographen-Drücker sind rechts so in das Gehäuse integriert, dass sie von vorn halb und von hinten ganz sichtbar sind. Die Uhr wird angetrieben von einem Kaliber 14 Automatik-Werk und ist auf der Brücke signiert.

Automatic new deep case design, screw back, brushed finish on top and larger polished sides. Black dial with two grey recessed subdials with orange 5-minute marks on the minute subdial and numbered hour indices from 1 to 12 on the hour register; they have concentric circular ridges.
The Autavia and Heuer logos are beneath 12 o'clock and; "GMT" is added below and "AUTOMATIC CHRONOGRAPH" is printed above the silver-framed date window at 6 o'clock. White-painted steel hands have luminous inserts. The bright yellow GMT hand completes a single revolution every 24 hours. The markers are filled with luminous material. Twin parallel markers are positioned at 12 o'clock and markers have been omitted at 3, 6, and 9 o'clock. The orange chronograph hand precisely reaches the orange hour markers at the outer 1/5-second track.
The matte black tension ring is fixed between the dial and the mineral glass. The red and blue, day and night, bezel has a luminous dot and is bidirectional, with a raster function denoting 24 hours. The crown is on the left side and signed "HEUER." The pushers appear integrated from the front but are fully visible from the back in contrast to the thinner case.
The watch is powered by a Calibre 14 automatic movement and is signed on the bridge.

AUTAVIA
HEUER
GMT
AUTOMATIC
CHRONOGRAPH
SWISS
G-BOAC

AUTAVIA

ref. 11630 MH

Finales Autavia-Gehäusedesign, kissenförmig, um die Bandanstöße vergrößert, mit verschraubtem Stahlboden, oben gebürstet und an den Seiten poliert. Kontrastreiches schwarzes Zifferblatt mit zwei mattgrauen vertieften Hilfszifferblättern, die konzentrische kreisförmige Rillen aufweisen. Weiße 5-Minuten-Markierungen auf dem Hilfszifferblatt, Indizes von 1, 2, 3 bis 12 auf dem Stunden-Hilfszifferblatt. Die Modellbezeichnung Autavia und das Heuer-Logo befinden sich unterhalb von 12 Uhr, der Schriftzug „AUTOMATIC CHRONOGRAPH“ ist über dem Datumsfenster bei 6 Uhr aufgedruckt.
Die weiß lackierten Zeiger haben große Leuchteinlagen. Die weißen Stahlindizes sind mit Leuchtmasse gefüllt. Ein größerer Index befindet sich bei 12 Uhr, die Indizes bei 3, 6 und 9 Uhr wurden weggelassen. Der weiße Chronographen-Zeiger erreicht präzise die weißen Stundenmarkierungen auf der äußeren 1/5-Sekunden-Skala.
Die mattschwarze Minuten- und Stunden-Lünette (MH) hat einen Leuchtpunkt, ist in beide Richtungen drehbar und verfügt über eine Rasterfunktion. Die Krone befindet sich auf der linken Seite und ist mit „HEUER“ signiert. Die geriffelten Drücker haben ein neues, ins Gehäuse integriertes, Design.
Die Uhr wird von einem Automatikwerk Kaliber 12 angetrieben und ist auf der Brücke mit „HEUER CAL.12“ signiert. Heuer entschied sich, an den in den Broschüren von 1982 und 1983 gezeigten Autavias „Jubiläums“-Armbänder zu verwenden.

Final Autavia case design, cushion style, enlarged around the lugs with steel screw back, brushed finish on top, and polished sides. Contrasting black dial with two matt grey recessed subdials, white 5-minute marks on minute subdial, indices from 1,2,3 to 12 on the hour register which evidence concentric circular ridges.
The Autavia name and Heuer logo are underneath 12 o'clock and "AUTOMATIC CHRONOGRAPH" is printed over the date window at 6 o'clock.
White-painted hands have large luminous inserts. White markers are filled with luminous material. A larger marker is positioned at 12 o'clock and marker have been omitted at 3, 6, and 9 o'clock. The white chronograph hand precisely reaches the white hour markers at the outer 1/5-second track.
The matt black minute and hour (MH) bezel has a luminous dot and is bidirectional, with a click function. The crown is on the left side and stamped "HEUER." The fluted pushers have a new integrated design.
The watch is powered by a Calibre 12 automatic movement and is signed on the bridge. Heuer chose to use "anniversary" bracelets on Autavias listed in the 1982 and 1983 catalogs.

Heue
Spezialul
Kollekti
1982
BLITZ
SERVIC
AUTAVIA
HEUER
AUTOMATIC
CHRONOGRAPH
SWISS

AUTAVIA

ref. 11630 „Diver 100“

Finales Autavia-Gehäusedesign, kissenförmig, um die Bandanstöße vergrößert, mit verschraubtem Stahlboden, gebürstetes Finish auf der Oberseite und polierte Seiten. Kontrastreiches schwarzes Zifferblatt mit bräunlichen Hilfszifferblättern, die konzentrische kreisförmige Rillen aufweisen. Weiße 5-Minuten-Markierungen auf dem Minuten-Hilfszifferblatt, Indizes von 1, 2, 3 bis 12 auf dem Stunden-Hilfszifferblatt.
Modellbezeichnung „AUTAVIA“ und Heuer-Logo unterhalb von 12 Uhr und über dem Datumsfenster bei 6 Uhr ist „AUTOMATIC CHRONOGRAPH“ aufgedruckt.
Die stählernen Mercedes-Zeiger haben Leuchteinlagen. Die runden weißen Indizes sind mit Leuchtmasse gefüllt. Ein dreieckiger Index mit voller Leuchtkraft befindet sich bei 12 Uhr, die Indizes bei 3, 6 und 9 Uhr wurden weggelassen. Der Chronographen-Zeiger aus Stahl hat einen integrierten Kreis, der mit Leuchtmasse gefüllt ist.
Die Taucher-Lünette mit schwarzen Minutenmarkierungen verfügt über einen Leuchtpunkt und eine Rasterfunktion. Die Krone befindet sich auf der linken Seite und ist mit „HEUER“ signiert. Die geriffelten Drücker haben ein neu gestaltetes, ins Gehäuse integriertes Design.
Die Uhr wird von einem Automatikwerk Kaliber 12 angetrieben und ist auf der Brücke „HEUER CAL.12“ signiert. Die Diver 100 wurde nur einmal, in einem Heuer-Katalog von 1982 mit Besonderheiten, aufgeführt, als Referenz 11630 Diver. Die Gehäuse der letzten Autavia-Serien trugen jedoch alle die Bezeichnung 11063V.

Final Autavia case design, cushion style, enlarged around the lugs with steel screw back, brushed finish on top, and polished sides. Contrasting black dial with brownish subdials, white 5-minute marks on minute subdial, markers from 1,2,3, to 12 on the hour register.
The Autavia name and Heuer logo are shown underneath 12 o'clock; "AUTOMATIC CHRONOGRAPH" is printed over the date window at 6 o'clock.
Steel Mercedes hands have luminous inserts. Round white markers are filled with luminous material. A full lume triangular index is positioned at 12 o'clock and indices have been omitted at 3, 6, and 9 o'clock. A steel chronograph hand has an integrated circle filled with luminous material.
The black minute diving bezel has a luminous dot plus a click function. The crown is on the left and signed "HEUER." The fluted pushers have a new integrated design.
The watch is powered by a Calibre 12 automatic movement and is signed on the bridge. The Diver 100 was only featured once, in a 1982 speciality Heuer catalog with the reference 11630 Diver; however, cases of the final Autavia series were all marked 11063V.

HEUER TIME GMBH
Service-Abteilung
8 München 19, Leonrodstr. 58, Tel
HEUER
Chro
Taucher 100
Flieger 10 000 m Höhe
AUTAVIA
HEUER
DIVER 100
SWISS

AUTAVIA
HEUER
AUTOMATIC
CHRONOGRAPH
SWISS

AUTAVIA

ref. 111.603

Olivfarbenes, beschichtetes Autavia-Gehäuse, kissenförmig, um die Bandanstöße vergrößert, mit verschraubtem Stahlboden. Olivfarbenes Zifferblatt mit zwei vertieften Hilfszifferblättern, die konzentrische kreisförmige Rillen aufweisen. Weiße 5-Minuten-Markierungen auf dem Hilfszifferblatt, Indizes von 1, 2, 3 bis 12 auf dem Stunden-Hilfszifferblatt.
Modellbezeichnung „Autavia" und Heuer-Logo unterhalb von 12 Uhr; Schriftzug „AUTOMATIC CHRONOGRAPH" über dem Datumsfenster bei 6 Uhr.
Die weiß lackierten Zeiger haben große Leuchteinlagen. Die Stahlindizes sind mit Leuchtmasse gefüllt. Ein größerer Index befindet sich bei 12 Uhr, die Indizes bei 3, 6 und 9 Uhr wurden weggelassen. Der weiße Chronographen-Zeiger reicht genau bis zu den weißen Stundenmarkierungen auf der äußeren 1/5-Sekunden-Skala.
Die olivbraune Minuten- und Stunden-Lünette (MH) hat einen Leuchtpunkt, ist in beide Richtungen drehbar und verfügt über eine Rasterfunktion. Die olivfarben beschichtete Krone befindet sich auf der linken Seite und ist mit „HEUER" signiert. Die olivfarben beschichteten, geriffelten Drücker haben ein neu gestaltetes, ins Gehäuse integriertes Design.
Die Uhr wird von einem Automatikwerk Kaliber 12 angetrieben und ist auf der Brücke mit „HEUER CAL.12" signiert. Hier ist ein olivfarben beschichtetes „Jubiläums"-Armband, wie im Heuer-Katalog von 1983 abgebildet, angebracht.

Olive-coated Autavia case, cushion style, enlarged around the lug, with steel screw back. Olive dial with two recessed with concentric circular ridges subdials, white 5-minute marks on minute subdial, indices from 1,2,3, to 12 on the hour register.
The Autavia name and the Heuer logo are underneath 12 o'clock and "AUTOMATIC CHRONOGRAPH" is printed over the date window at 6 o'clock.
White-painted hands have large luminous inserts. Steel markers are filled with luminous material. A larger marker is positioned at 12 o'clock and markers have been omitted at 3, 6, and 9 o'clock. The white chronograph hand precisely reaches the white hour markers at the outer 1/5 second track.
The olive-brown minute and hour (MH) bezel has a luminous dot and is bidirectional with a click function. The olive-coated crown is on the left and signed "HEUER." The olive coated fluted pushers have a new integrated design.
The watch is powered by a Calibre 12 automatic movement and is signed on the bridge. An olive-coated anniversary bracelet as shown in the 1983 Heuer catalog is attached.

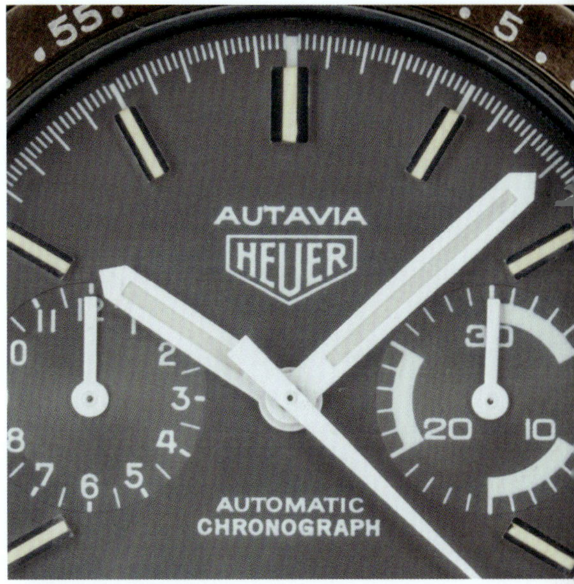

AUTAVIA

ref. 113.603 „Military IDF"

Schwarz beschichtetes Autavia-Gehäuse, kissenförmig, um die Bandanstöße vergrößert, mit verschraubtem Stahlboden. Schwarzes Zifferblatt mit zwei anthrazitfarbenen vertieften Hilfszifferblättern, die konzentrische kreisförmige Rillen aufweisen. Weiße 5-Minuten-Markierungen auf dem Hilfszifferblatt, Indizes von 1, 2, 3 bis 12 auf dem Stunden-Hilfszifferblatt.
Modellbezeichnung „AUTAVIA" und Heuer-Logo unterhalb von 12 Uhr und der Schriftzug „AUTOMATIC CHRONOGRAPH" über dem Datumsfenster bei 6 Uhr.
Die weiß lackierten Zeiger haben große Leuchteinlagen. Die weißen Stahlindizes sind mit Leuchtmasse gefüllt. Ein größerer Index befindet sich bei 12 Uhr, die Indizes bei 3, 6 und 9 Uhr wurden weggelassen. Der weiße Chronographen-Zeiger erreicht die weißen Stundenmarkierungen genau auf der äußeren 1/5-Sekunden-Skala.
Die schwarze Minuten- und Stunden-Lünette (MH) ist in beide Richtungen drehbar und verfügt über eine Rasterfunktion. Die schwarz beschichtete Krone befindet sich auf der linken Seite und ist mit „HEUER" signiert. Die schwarz beschichteten, geriffelten Drücker haben ein neues, im Gehäuse integriertes Design.
Die Uhr wird von einem Automatikwerk Kaliber 12 angetrieben und ist auf der Brücke mit „HEUER CAL.12" signiert. Hier ist ein schwarz beschichtetes „Jubiläums"-Armband, wie im Heuer-Katalog von 1983 abgebildet, angebracht. Der äußere Gehäuseboden trägt militärische Markierungen der Israel Defense Force (IDF).

Black-coated Autavia case, cushion style, enlarged around the lugs, steel screw back. Black dial with charcoal-grey recessed subdials with concentric circular ridges, white 5-minute marks on minute subdial, markers from 1,2,3 to 12 on the hour register.
The "AUTAVIA" name and Heuer logo are underneath 12 o'clock and "AUTOMATIC CHRONOGRAPH" is printed over the date window at 6 o'clock.
The white-painted hands have large luminous inserts. White markers are filled with luminous material. A larger marker is positioned at 12 o'clock and markers have been omitted at 3, 6, and 9 o'clock. The white chronograph hand precisely reaches the white hour markers at the outer 1/5-second track.
The black minute and hour (MH) bezel is bidirectional with a click function. The black-coated crown is on the left and is signed "HEUER." The black-coated fluted pushers have a new integrated design.
The watch is powered by a Calibre 12 automatic movement and is signed on the bridge. A black-coated anniversary bracelet as shown in the 1983 Heuer catalog is attached. The outer case back shows military markings of the Israel Defense Force (IDF).

AUTAVIA
HEUER
AUTOMATIC
CHRONOGRAPH
SWISS
18

AUTAVIA
HEUER
AUTOMATIC
CHRONOGRAPH
SWISS
18

AUTAVIA

ref. 114.603

Vergoldetes Autavia-Gehäuse, kissenförmig, um die Bandanstöße vergrößert, mit verschraubtem Stahlboden. Goldfarbiges Zifferblatt mit zwei vertieften Hilfszifferblättern, die konzentrische kreisförmige Rillen aufweisen. Schwarze 5-Minuten-Markierungen auf dem Hilfszifferblatt, Indizes von 1, 2, 3 bis 12 auf dem Stunden-Hilfszifferblatt.
Modellbezeichnung „AUTAVIA" und Heuer-Logo sind unterhalb von 12 Uhr und „AUTOMATIC CHRONOGRAPH" ist über dem Datumsfenster bei 6 Uhr aufgedruckt.
Die schwarz lackierten Zeiger haben große Leuchteinlagen. Die goldenen Indizes sind mit Leuchtmasse gefüllt. Ein größerer Index befindet sich bei 12 Uhr, die Indizes bei 3, 6 und 9 Uhr wurden weggelassen. Der schwarze Chronographen-Zeiger erreicht präzise die schwarzen Stundenmarkierungen auf der äußeren 1/5-Sekunden-Skala.
Die schwarze Minuten- und Stunden-Lünette (MH) mit Leuchtpunkt ist in beide Richtungen drehbar und verfügt über eine Rasterfunktion. Die vergoldete Krone befindet sich auf der linken Seite und ist mit „HEUER" signiert. Die vergoldeten, geriffelten Drücker haben ein neues, ins Gehäuse integriertes Design.
Die Uhr wird von einem Automatikwerk Kaliber 12 angetrieben und ist auf der Brücke „HEUER CAL.12" signiert. Hier ist ein vergoldetes, beschichtetes „Jubiläums"-Armband, wie im Heuer-Katalog von 1983 abgebildet, angebracht.

Gold-plated Autavia case, cushion style, enlarged around the lugs, with steel screw back. Golden dial with two recessed subdials with concentric circular ridges, black 5-minute marks on minute subdial, markers from 1, 2, 3 to 12 on the hour register.
The Autavia name and Heuer logo are shown underneath 12 o'clock; "AUTOMATIC CHRONOGRAPH" is printed over the date window at 6 o'clock.
Black-painted hands have large luminous inserts. Golden markers are filled with luminous material. A larger marker is positioned at 12 o'clock and markers have been omitted at 3, 6, and 9 o'clock. The black chronograph hand precisely reaches the black hour markers at the outer 1/5-second track.
The black minute and hour (MH) bezel with a luminous dot is bidirectional and has a raster function. The gold-plated crown is on the left and is signed "HEUER." The gold-plated fluted pushers have a new integrated design.
The watch is powered by a Calibre 12 automatic movement and is signed on the bridge. A gold-plated coated anniversary bracelet as shown in the 1983 Heuer catalog is attached.

Examples of speed conversions

135 km/h = ? km/min = ? km/s?

As 1 h = 60 min = 3,600 s, 135 km/h $= \frac{135 \text{ km}}{60 \text{ min}} = \frac{135 \text{ km}}{3{,}600 \text{ s}}$

Find the number 1.35 on the outer ring and set it opposite to the number 3.6 on the inner scale. 3.75 will be indicated above the mark ▲ and 2.25 opposite to the number 6. Therefore 135 km/h = 2.25 km/min = 37.5 m/s.

250 mph = ? km/h = ? knots?
Set the number 2.5 on the outer ring opposite to the number 6 (mph). 4.02 will be indicated on the outer ring opposite to the 'km' dot and 2.16 opposite to the 'kt' dot. Therefore 250 mph = 402 km/h = 216 knots.

CALCULATOR

1972 erschien ein neues Modell mit dem mathematisch klingenden Namen „Calculator“. Jeder Calculator wurde mit einem kleinen Buch ausgeliefert, in dem alle verschiedenen Rechenoperationen genauestens erklärt werden:
Der Heuer Calculator wurde ab 1972 mit zwei verschiedenen Automatikuhrwerken angeboten – sowohl mit einem Kaliber 12 als auch mit einem um zehn Prozent günstigeren Kaliber 15. Das Design des Kaliber 12 war klassisch und besaß ein in 12 Stunden und 30 Minuten unterteiltes Hilfszifferblatt. Bei Kaliber 15 entfiel die 12-Stunden-Anzeige zu Gunsten einer kleinen permanenten Sekunde auf 10 Uhr. Zwei Jahre später ergänzte eine Version zum manuellen Aufziehen mit der Referenz 740.633 das Portfolio. Auf dem blauen Zifferblatt fehlte der übliche Schriftzug „AUTOMATIC CHRONOGRAPH“. Die Handaufzugsvariante war außerdem durch die Position der Krone zwischen den Drückern auf 3 Uhr gut zu erkennen. Diese Variante sowie die 1977 erschienene multifunktionelle Automatikversion mit der Referenz 11630 MHT, ohne „CALCULATOR“ auf dem schwarzen Zifferblatt, zählt heute zu den seltensten Heuer-Uhren.
Mit dem Heuer Calculator hatte der Wissende ein mathematisch ausgeklügeltes Instrument an der Hand, mit dem er durch wenige Griffe die Resultate schwieriger Rechenoperationen an der Uhr einfach ablesen konnte.
Calculator Chronographen (Caliber 12 oder 15) hatten blaue oder schwarze Zifferblätter.
Der Heuer Calculator hatte einen langen Produktlebenszyklus von 1972 bis zum Übergang zu TAG Heuer Mitte der 80er Jahre.

In 1972 Heuer presented a new model line with the mathematical name of "Calculator." Every Calculator was sold with a booklet containing precise explanations of all mathematical calculations.
From 1972, the Heuer Calculator was available with two different automatic movements—one a Calibre 12, and one a Calibre 15 movement that was ten percent cheaper. The Calibre 12 had a classic design with a 12-hour and a 30-minute subdial. The Calibre 15 omitted the 12-hour display in favour of a small permanent second at 10 o'clock. Two years later, a manual winding version with the reference 740.633 was added to the portfolio. The blue dial differed in its lack of the "AUTOMATIC CHRONOGRAPH" print. The manual winding version was also clearly identifiable by the crown's position between the pushers at 3 o'clock. The manual winding version, along with the multifunctional automatic version released in 1977 (lacking the "CALCULATOR" text on the black dial) and reference 11630 MHT, are now regarded as some of the rarest Heuer watches.
With the Heuer Calculator, the wearer had a clever mathematical instrument at hand which allowed complex mathematical calculations to be performed with just a few clicks of the watch.
Calculator chronographs (Calibre 12 or 15) had blue or black faces. The Heuer Calculator had a long life, from its launch in 1972 until the company became TAG Heuer in the mid-1980s.

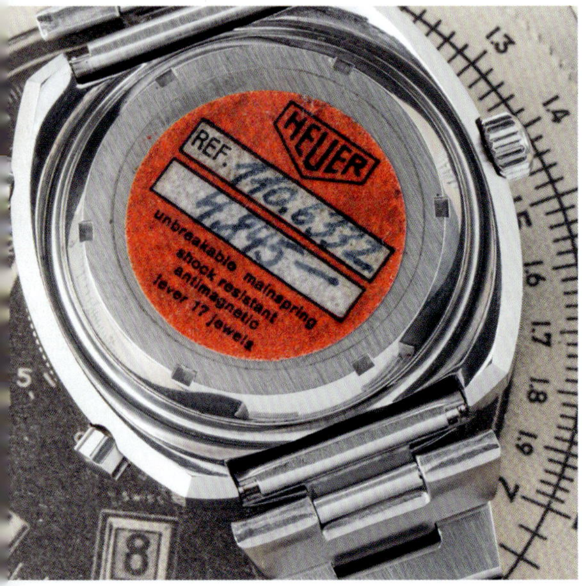

CALCULATOR

ref. 110.633

Rechteckiges Gehäuse mit verschraubtem Boden, Bürstenschliff oben und seitlich sowie zwei Rechenkreisen, die aus einer fixen inneren Skala und einem rotierenden äußeren Ring bestehen. Das tiefblaue Zifferblatt hat 30-Minuten- und 12-Stunden-Hilfszifferblätter und ist umgeben von einem polierten Spannring. Das Minuten-Hilfszifferblatt hat drei orangefarbene 5-Minuten-Markierungen. Der Modellname „CALCULATOR" steht unter dem Heuer-Logo, „AUTOMATIC CHRONOGRAPH" über dem stahlgerahmten Datumsfenster bei 6 Uhr.
Stundenindizes aus Stahl sind mit Leuchtmaterial gefüllt und am äußeren Rand platziert. Ein größerer Index steht bei 12 Uhr, die Indizes bei 3, 6, und 9 Uhr fehlen. Orangefarbene Zeiger haben große Leuchtmaterial-Einschlüsse. Der orangefarbene Chronographen-Sekundenzeiger hat ein Leuchtmaterial-Rechteck und erreicht die orangefarbenen Stunden-Indizes der weißen 1/5-Sekunden-Skala. Die Krone ist auf der linken Seite und mit „HEUER" signiert. Die Chronographen-Drücker sind rechts integriert. Diese Uhr wurde vorbereitet für ein Handaufzugswerk, deshalb sieht man ein Bohrloch auf der rechten Seite zwischen den Drückern. Der Rückdeckel hat einen Längsschliff mit polierten Seiten.
Das Uhrwerk ist ein Kaliber 12 und ist auf der Brücke mit „CAL. 12" signiert.

Rectangular screw-back case has brushed finish on top and sides and two calculating circles consisting of a fixed inner scale and a rotating outer ring. The deep blue dial has 30-minute and 12-hour recording subdials and is surrounded by a polished tension ring. The minute subdial has three orange circular markings indicating 5 minutes each. The model name "CALCULATOR" is positioned below the Heuer logo under 12 o'clock. "AUTOMATIC CHRONOGRAPH" is printed above the steel-framed date window at 6 o'clock.
Steel hour markers filled with luminous materials are placed at the outer end of the deep blue dial. A larger marker is at 12 o'clock and markers have been omitted at 3, 6, and 9 o'clock. Orange-painted hands have large luminous inserts. The orange chronograph seconds hand shows a luminous rectangle and reaches the orange hour indices on the white 1/5-second track. The crown is integrated on the left and signed "HEUER." Chronograph pushers are integrated on the right. This watch case was prepared for a manual wind movement, hence a drill-hole can be seen between the pushers on the right side. The case back has longitudinal grinding with polished sides.
The movement is a Calibre 12, and is signed "CAL. 12" on the bridge.

HEUER
AUTOMATIC
CHRONOGRAPH
SWISS

HEUER
AUTOMATIC
CHRONOGRAPH
SWISS
TACHY
15 PULSATIONS

CALCULATOR

ref. 11633 MHT

Rechteckiges Gehäuse mit verschraubtem Boden, Bürstenschliff oben und seitlich sowie zwei Rechenkreisen, die aus einer fixen inneren Skala und einem rotierenden äußeren Ring bestehen. Das schwarze Zifferblatt hat 30-Minuten- und 12-Stunden-Hilfszifferblätter und ist umgeben von einem polierten Spannring. Das Minuten-Hilfszifferblatt hat drei weiße 5-Minuten-Markierungen. Der Modellname „CALCULATOR" fehlt bei diesem Minuten-Stunden-Tachymeter (MHT) Chronograph mit Pulsometer-Skala. Das Heuer-Logo steht unter 12 Uhr, „AUTOMATIC CHRONOGRAPH" über dem weißgerahmten Datumsfenster bei 6 Uhr.
Die ganz mit Leuchtmasse gefüllten Stundenindizes sind am äußeren Rand des schwarzen Zifferblattes platziert, kleinere Indizes bei 3, 6, und 9 Uhr. Weiße Zeiger haben große Leuchtmaterial-Einschlüsse. Der weiße Chronographen-Sekundenzeiger erreicht die weiße 1/5-Sekunden-Skala. Die Krone ist auf der linken Seite und mit „HEUER" signiert. Die Chronographen-Drücker sind rechts integriert. Der Rückdeckel hat einen Längsschliff mit polierten Seiten. Das Uhrwerk ist ein Kaliber 12 und ist mit „CAL. 12" auf der Brücke signiert.

Rectangular screw-back case has a brushed finish on top and sides and two calculating circles consisting of a fixed inner scale and a rotating outer bezel. The black dial features 30-minute and 12-hour recording subdials and is surrounded by a polished tension ring. The minute subdial has three white circular markings indicating 5 minutes each. The model name "CALCULATOR" was omitted on this "minute, hour, tachymeter" (MHT) chronograph with pulsometer scale. The Heuer logo remains below 12 o'clock. "AUTOMATIC CHRONOGRAPH" is printed above the white-framed date window at 6 o'clock.
All luminous hour markers are incorporated on the outer edge of the black dial with smaller markers at 3, 6, and 9 o'clock. The white painted hands have large luminous inserts. The white chronograph hand reaches the white 1/5-second track on the outer edge of the dial. The crown is integrated on the left and signed "HEUER." Chronograph pushers are integrated on the right. The case back has longitudinal grinding with polished sides.
The movement is a Calibre 12, and is signed "CAL. 12" on the bridge.

HEUER
CALCULATOR
AUTOMATIC
CHRONOGRAPH
SWISS

CALCULATOR

ref. 150.633

Rechteckiges Gehäuse mit verschraubtem Boden, Bürstenschliff oben und seitlich sowie zwei Rechenkreisen, die aus einer fixen inneren Skala (mit seltenem Rotdruck) und einem rotierenden äußeren Ring bestehen. Das Zifferblatt ist schwarz mit einer kleinen permanenten Sekundenanzeige bei 10 Uhr und einem 30-Minuten-Hilfszifferblatt mit orangefarbenen 5-Minuten-Markierungen. Der Modellname „CALCULATOR" steht bei 9 Uhr, das Heuer-Logo unter 12 Uhr und „AUTOMATIC CHRONOGRAPH" über dem stahlgerahmten Datumsfenster bei 6 Uhr. Stundenindizes aus Stahl sind mit Leuchtmaterial gefüllt und am äußeren Rand des schwarzen Zifferblattes platziert. Ein größerer Index steht bei 12 Uhr, die Indizes bei 3 und 6 Uhr fehlen.
Orangefarbene Zeiger haben große Leuchtmaterial-Einschlüsse. Der orangefarbene Chronographen-Sekundenzeiger hat ein Leuchtmaterial-Rechteck und erreicht die orangefarbenen Stundenindizes der weißen 1/5-Sekunden-Skala. Die Krone ist auf der linken Seite integriert und mit „HEUER" signiert. Die Drücker sind rechts integriert. Der Rückdeckel hat einen Längsschliff mit polierten Seiten. Das Uhrwerk ist ein Kaliber 15 und ist mit „CAL. 15" auf der Brücke signiert.

Rectangular screw-back case with brushed finish on top and sides and two calculating circles consisting of a fixed inner scale with rare red print and a rotating outer ring. The dial is black with a small permanent white second hand at 10 o'clock and a 30-minute recording subdial with three orange circular markings indicating 5 minutes each.
The model name "CALCULATOR" is at 9 o'clock while the Heuer logo is below 12 o'clock and "AUTOMATIC CHRONOGRAPH" is printed above the steel-framed date window at 6 o'clock. Steel hour indices filled with luminous material are placed at the outer end of the black dial. A larger index is at 12 o'clock and indices have been omitted at 3 and 6 o'clock.
Orange painted hands have large luminous inserts. The orange chronograph second hand shows a luminous rectangle and reaches the orange hour markers on the white 1/5-second track. The crown is integrated on the left and "HEUER" signed. Chronograph pushers are integrated on the right. The case back has longitudinal grinding with polished sides.
The movement is a Calibre 15, and is signed "CAL.15" on the bridge.

Ref. 110.633 Calcula

Mit dem Präzisions-Rechensch ti-
plikationen, Divisionen, Umwand
einer Abweichung des Messresult
als 1% möglich. Ideal für Ingenieu
Wissenschafter. Edelstahlgehäuse, w

Ref. 740.633 Calculator

CALCULATOR

ref. 740.633

Rechteckiges, verschraubtes Gehäuse, Bürstenschliff oben und seitlich sowie zwei Rechenkreise, die aus einer fixen inneren Skala und einem rotierenden äußeren Ring bestehen. Im tiefblauen Zifferblatt sitzen zwei Hilfszifferblätter, das eine mit 30-Minuten-Countdown, das andere mit 12-Stunden-Skala, und ist umgeben von einem polierten Spannring. Das Minuten-Hilfszifferblatt besitzt drei orangefarbene 5-Minuten-Markierungen. Unter dem Heuer-Logo steht der Modellname „CALCULATOR", über dem stahlgerahmten Datumsfenster auf 6 Uhr steht „SWISS".
Die Stundenindizes aus Stahl sind mit Leuchtmaterial gefüllt und am äußeren Rand platziert, die Indizes bei 3, 6, und 9 Uhr fehlen. In die orangefarbenen Zeiger ist Leuchtmaterial eingeschlossen. Der orange Chronographen-Sekundenzeiger hat ein Rechteck mit Leuchtmaterial an der Spitze, welche die orangefarbenen Stundenindizes der weißen 1/5-Sekunden-Skala erreicht. Die Krone ist mit „HEUER" signiert und bei der Handaufzugsversion zwischen den Chronographen-Drückern rechts integriert. Der Rückdeckel hat einen Längsschliff mit polierten Seiten. Angetrieben wird die Uhr von einem Valjoux 7740, auf der Brücke ist sie mit „Heuer CAL. 7740" signiert.

Rectangular screw-back case with brushed finish on top and sides and two calculating circles consisting of a fixed inner scale and a rotating outer bezel. The deep blue dial has 30-minute and 12-hour recording subdials and is surrounded by a polished tension ring. The minute subdial has three orange circular markings indicating 5 minutes each. The model name "CALCULATOR" is positioned below the Heuer logo and "SWISS" is printed above the steel-framed date window above 6 o'clock.
Steel hour markers filled with luminous materials are placed at the outer edge of the dial and markers have been omitted at 3, 6, and 9 o'clock.
The orange-painted hands have large luminous inserts. The orange chronograph second hand shows a luminous rectangle and reaches the orange hour markers on the white 1/5-second track.
The crown is "HEUER" signed and integrated between the pushers on the right for the manual winding version. The case back has longitudinal grinding with polished sides.
The movement is a Valjoux 7740 and is signed "HEUER CAL.7740" on the bridge.

Der Chevrolet Camaro war das Indianapolis „Pace Car" 1967 und 1969.

The Chevrolet Camaro was the pace car in Indianapolis in 1967 and 1969.

CAMARO

Der Camaro war ein außergewöhnliches Auto, er hatte ein modernes Design, eine stromlinienförmige Vorderfront, die „Coca-Cola“-Linie unterstrich die klassische Coupé-Variante und führte zum prägnant kurzen Heck.
Die Camaro Uhr glänzt durch den Strahlenschliff auf dem rechteckigen Gehäuse mit konkav geschwungenen Seiten.
Der Camaro war Chevrolets Antwort auf den sehr erfolgreichen Mustang von Ford.
Die Camaro war eine völlig neue Produktlinie mit Design von Heuer.
Der Camaro von Chevrolet sollte ein Icon der neuen Pony-Cars werden (kleiner, leichter und schneller als die großen „Schlitten“ der älteren Generation) und eine neue jüngere Zielgruppe erobern. Der Name „CAMARO“ kommt aus dem Französischen, bedeutet „Freund, Wegbegleiter, ein guter Kamerad“ und war in den USA sofort erfolgreich, da er von Sportwagenkäufern als positiv wahrgenommen und mit Qualität verknüpft wurde. Die Markteinführung erfolgte am 26. September 1966, gerade zur richtigen Zeit, um eine maßgebliche Rolle in der amerikanischen Muscle-Car-Szene zu spielen. Starke Motorisierung in bezahlbaren Mittelklasse-Coupés war der große Trend.
1967 und 1969 wurde der Camaro „Super Sport“ zum offiziellen „Pace Car“ des „Indianapolis 500“, des wohl berühmtesten Autorennens auf amerikanischem Boden. Ein „Pace Car“ hat, wie heute in der Formel 1, die verantwortungsvolle Aufgabe, das Starterfeld anzuführen und nach Unfällen oder Rennunterbrechungen das Tempo vorzugeben. Mit diesem Teil der Rennwelt war eine enorme Medienpräsenz verbunden, was den Wiedererkennungswert des Heuer-Logos massiv steigerte und bei der Einführung der sportlichen Camaro-Armbanduhr von großem Vorteil war.

The Camaro was a very special automobile; it had a modern new design, a streamlined front, and a "Coca-Cola" line that accentuated the classic coupé and led to the striking short tail.
The Camaro wristwatch features a gleaming, polished grind on its rectangular case and curved sides.
The Camaro automobile was Chevrolet's answer to the exceptionally successful Ford Mustang.
The Camaro wristwatch was a totally new product line designed by Heuer.
The Camaro automobile was to become an icon among the new "pony cars" (smaller, lighter, and faster than the large cars of the older generation) and win over a new, younger group of customers. The name "CAMARO" comes from the French and means "friend, companion, good guy." The name was an immediate hit in the United States, quickly accepted by sports car buyers and appreciated as an expression of value and quality. It was launched on September 26, just in time to assume an important role on the American muscle car scene. At that time a powerful engine in an affordable mid-range car was a popular combination.
1967 to 1969, the Camaro Super Sport was the official pace car of the Indianapolis 500, the most famous racing competition in the USA. The pace car is responsible for leading leading the field at the start and setting the pace after accidents or race interruptions, as in today's Formula 1. This element of the racing world delivered an enormous media presence for the Heuer product line and resulted in widespread logo recognition following the introduction of the sporty Camaro wristwatch. Jack Heuer was keenly interested in the US market. Now he could break into the motorsport fan market with his new Camaro model line and develop

Jack Lemmon trägt eine Camaro.

Jack Lemmon wearing a Camaro.

Der amerikanische Markt lag Jack Heuer besonders am Herzen, jetzt konnte er mit dem neuen Camaro-Modell in diese junge, sehr am Motorsport interessierte Zielgruppe vordringen und dabei viele Synergien nutzen. Die Kernkompetenz der Zeitmessung bei großen Motorsport-Veranstaltungen half, den Namen Heuer in der Öffentlichkeit hervorragend zu positionieren.

Die Modellreihe Camaro bestand aus Chronographen mit zwei und drei Hilfszifferblättern, angetrieben durch Handaufzugswerke aus dem Hause Valjoux. Die Gehäuse waren meist aus Stahl, einige vergoldet mit Stahlboden. Die 18-Karat-Version ist extrem selten. Die Camaro blieb bis 1972 in den Katalogen und lief dann aus.

new synergies. His core competence in timing great motorsport events helped clothe the name Heuer in excellent publicity.

The Camaro wristwatch line comprised chronographs with 2 or 3 subregisters, driven by Valjoux manual movements. The cases were mostly steel, some gold-plated on a steel ground.

The 18-karat version of this wristwatch is extremely rare. The Camaro wristwatch appeared in catalogs until 1972, after which it faded away.

F
Nikon

Der südafrikanische Tyrrell-Fahrer Jody Scheckter mit Heuer Camaro, 1974.

South African Tyrrell driver Jody Scheckter with Heuer Camaro in 1974.

F
Nikon

TACHYMETER
CAMARO
HEUER
SWISS

CAMARO

ref. 7220 NT

Kissenförmiges Gehäuse mit gebürsteter Oberfläche. Mattschwarzes Zifferblatt mit glänzend schwarzen Hilfszifferblättern und spektakulärer roter Tachymeter-Skala. Die 30-Minuten- und 12-Stunden-Hilfszifferblätter mit roten Zeigern stehen bei 3 bzw. 6 Uhr. Das Sekunden-Hilfszifferblatt ist bei 9 Uhr. Die Hilfszifferblätter sind versenkt und zeigen konzentrische Rillen. Die Stundenindizes sind aus Metall, der Index bei 12 Uhr ist größer als die anderen, sie haben weiße Einlagen und kleine Leuchtpunkte am äußeren Ende. „CAMARO" steht über dem Heuer-Logo. Polierte Stahlzeiger haben schwarze und Leuchtmasse-Einschlüsse. Der Minutenzeiger ist verkürzt und steigert dadurch die Lesbarkeit der Tachymeter-Skala am Rande des Zifferblattes. Der rote Chronographen-Sekundenzeiger und die kleinen roten Zeiger der Hilfszifferblätter sind klar lesbar. Die Krone ist mit „HEUER" signiert. Die Drücker sind asymmetrisch und von hinten sichtbar. Der Boden der Camaro ist verschraubt. Die Innenseite des Bodens ist mit „HEUER-LEONIDAS SA" signiert.
Das Uhrwerk ist ein Valjoux 72, signiert mit „72 R" unter der Unruh und mit „HEUER-LEONIDAS SA" auf der Brücke.

Cushion-shaped case with brushed finish on top. Matte black dial, glossy black subdials and spectacular red tachymeter scale. The 30-minute and 12-hour totalizers with red hands are positioned at 3 and 6 o'clock. The second subdial is at 9 o'clock. The subdials are recessed and have concentric circular ridges.
The hour markers are in metal, the 12 being larger than the rest, with white inserts and small luminous dots at the outer end. "CAMARO" is positioned above the Heuer logo. The polished steel hands have black and luminous inserts. The minute hand is shortened for increased legibility of the tachymeter scale at the edge of the dial. The red chronograph second hand and small red hands of the chronograph subdials are clearly legible. The crown is signed "HEUER." The chronograph pushers are asymmetric and visible from the back. The Camarohas a screw back; the inner back case is signed "HEUER-LEONIDAS SA."
Powered by a Valjoux 72 movement signed "72 R" under the balance wheel and "HEUER-LEONIDAS SA" on the bridge.

Der südafrikanische Tyrrell-Fahrer Jody Scheckter mit Heuer Camaro, 1974.

South African Tyrrell driver Jody Scheckter with Heuer Camaro in 1974.

CAMARO
HEUER
SWISS
7228 CH
World wide service
Service mondial
Servicio mundial
Internationaler Service

CAMARO

ref. 7228 CH

Kissenförmiges 18-Karat-Goldgehäuse mit gebürsteter Oberfläche. Das goldene Zifferblatt hat drei goldene Hilfszifferblätter. Die 30-Minuten- und 12-Stunden-Hilfszifferblätter stehen bei 3 bzw. 6 Uhr, das Sekunden-Hilfszifferblatt bei 9 Uhr. Die Hilfszifferblätter sind versenkt und zeigen konzentrische Rillen.
Die Stundenindizes sind golden, der Index bei 12 Uhr ist größer als die anderen, sie haben Bürstenschliff, schräge Facetten und kleine Leuchtpunkte am äußeren Ende. „CAMARO“ steht über dem Heuer-Logo. Polierte Goldzeiger haben schwarze und Leuchtmasse-Einlagen. Der schwarze Chronographen-Sekundenzeiger und die kleinen schwarzen Zeiger der Chronographen-Hilfszifferblätter sind klar lesbar.
Die Krone ist mit „HEUER“ signiert. Die Chronographen-Drücker sind asymmetrisch und von hinten sichtbar. Die Punzierung liegt zwischen den Bandanstößen. Das Camaro-Gehäuse hat einen verschraubten Boden mit polierter Oberfläche. Die Innenseite des Gehäuses ist mit „HEUER-LEONIDAS SA 18 K“ signiert. Das Uhrwerk ist ein Valjoux 72, signiert mit „72 R“ unter der Unruhe und mit „HEUER-LEONIDAS SA“ auf der Brücke. Die Boden-Innenseite zeigt die Goldmarkierung und hat einen Ringelschliff.

Cushion-shaped 18-karat gold case has brushed finish on top. Gold dial with gold subdials. The 30-minute and 12-hour totalizers are positioned at 3 and 6 o'clock. The second subdial is at 9 o'clock. The subdials are recessed with concentric circular ridges.
The hour markers are golden, the 12 being larger than the rest, with brushed finish, angled facets, and small luminous dots at the outer end. "CAMARO" is positioned above the Heuer logo. The polished gold hands have black and luminous inserts. The black chronograph second hand and small black hands of the chronograph subdials are clearly legible.
The crown is signed "HEUER." The chronograph pushers are asymmetric and visible from the back. The hallmark is between the lugs. The Camaro has a screw back with a polished finish; the inner back case is signed "HEUER-LEONIDAS SA 18 K."
Powered by a Valjoux 72 movement signed "72 R" under the balance wheel and "HEUER-LEONIDAS SA" on the bridge. The inner case displays the gold hallmarks and has ring grinding.

CAMARO
HEUER
SWISS
T

TACHYMETER
CAMARO
HEUER
SWISS
3042-01

CAMARO

ref. 73643 T „Panda“

Leicht vergrößertes, kissenförmiges Camaro-Gehäuse mit gebürsteter Oberseite. Weißes Zifferblatt mit kontrastierenden grauschwarzen Hilfszifferblättern und passender Tachymeter-Skala. Die 30-Minuten- und 12-Stunden-Totalisatoren mit roten Zeigern befinden sich bei 3 und 6 Uhr. Eine permanente laufende Sekunde mit weißem Zeiger ist bei 9 Uhr positioniert. Die drei Hilfszifferblätter sind in konzentrischen Kreisen in das Zifferblatt eingelassen.
Die neu gestalteten Stundenindizes sind aus Metall (der auf 12 Uhr ist größer als die übrigen), mit kleinen Leuchtpunkten am äußeren Ende. Der Modellname „CAMARO“ ist über dem Heuer-Logo positioniert. In die Zeiger aus poliertem Stahl mit schwarzen Linien ist eine breitere Leuchtmasse integriert. Der rot lackierte Chronographen-Sekundenzeiger sowie die kleinen roten Zeiger auf den Minuten- und Stunden-Zifferblättern sind sehr gut ablesbar.
Die Krone ist mit „HEUER“ signiert. Die Chronographen-Drücker sind asymmetrisch und von der Rückseite aus vollständig sichtbar. Die Camaro hat einen verschraubten Gehäuseboden mit gebürstetem Längsschliff, dessen Innenseite mit „HEUER-LEONIDAS SA“ signiert ist.
Angetrieben wird sie von einem Valjoux 7736-Werk mit der Signatur „7736 R“ unter der Unruh und „HEUER LEONIDAS SA“ auf der Brücke. Die Valjoux 7736-Werke lösten 1971 sukzessive die Uhrwerke Valjoux 72 ab. Bereits zwei Jahre später verschwand die Camaro aus den Heuer-Preislisten.

Slightly enlarged Camaro cushion-shaped case with brushed finish on top. White dial with contrasting charcoal subdials and matching tachymeter scale. The 30-minute and 12-hour totalisers with red hands are positioned at 3 and 6 o'clock. A permanent running second subdial have circular ridges.
The new-design hour markers are in metal, the one at 12 being larger than the rest, with black inserts and small luminous dots at the outer end. The "CAMARO" model name is positioned over the Heuer logo. Polished steel hands have widened luminous inlays and black lines. The red-painted chronograph second hand as well as the small red hands on the minute and hour subdials are clearly legible.
The Crown is signed "HEUER." The chronograph pushers are asymmetric and fully visible from the back. The Camaro has a screw back with a longitudinal brushed finish: the interior of the case is signed "HEUER-LEONIDAS SA."
Powered by a Valjoux 7736 movement signed "7736 R" under the balance wheel and "HEUER-LEONIDAS SA" on the bridge. The Valjoux 7736 movement succeeded the Valjoux 72 in 1971. The Camaro disappeared from Heuer price lists two years later.

CAMARO
HEUER
TACHYMETER
SWISS
18

CAMARO

ref. 73445 T

Vergoldetes, kissenförmiges Gehäuse mit gebürsteter Oberfläche. Champagnerfarbenes Zifferblatt und zwei Hilfszifferblätter mit Tachymeter-Skala. Die permanente Sekunde und das 30-Minuten-Hilfszifferblatt befinden sich bei 9 und 3 Uhr. Die Hilfszifferblätter sind in das Zifferblatt eingelassen und weisen kreisförmige Rillen auf. Bei 6 Uhr ist eine Datumsanzeige zu sehen. Die neu gestalteten Stundenindizes sind golden mit schwarzen Einlagen, der 12-Uhr-Index ist vergrößert. Alle Indizes haben kleine Leuchtpunkte am äußeren Ende. „CAMARO" ist über dem Heuer-Logo positioniert.
Die polierten goldfarbenen Zeiger haben schwarze Einsätze und integrierte Leuchtmasse. Der schwarze Chronographen-Sekundenzeiger und die kleinen schwarzen Hilfszifferblattzeiger sind gut ablesbar.
Drücker und Krone mit Heuer-Signatur sind vergoldet. Die Camaro hat einen verschraubten Stahlboden mit gebürstetem Längsschliff, das Innere des Gehäuses ist mit „HEUER-LEONIDAS SA" signiert. Angetrieben wird sie von einem Valjoux-7734 mit der Signatur „7734 R" unter der Unruh sowie „HEUER LEONIDAS SA" auf der Brücke. Diese Camaro weist ein seltenes, vergoldetes Armband mit signierter Heuer-Schließe auf

Gold-plated cushion-shaped case with brushed finish on top. Champagne dial and subdials with tachymeter scale. The permanent second and 30-minute subdials are at 9 and 3 o'clock. The subdials are recessed in the dial and have circular ridges. A date display is incorporated at the 6 o'clock position.
The new-design hour markers are in gold with black inlays. The 12 o'clock marker is larger, all markers have small luminous dots at the outer bevel. "CAMARO" is positioned over the Heuer logo. The polished gold hands have black inserts and integrated luminous material.
The black chronograph second-hand and small black subdial hands are clearly legible.
The pushers and the Heuer signed crown are gold-plated. The Camaro has a steel screw back with longitudinal brushed finish; the interior of the case is signed "HEUER-LEONIDAS SA." Powered by a Valjoux 7734 movement signed "7734 R" under the balance wheel and "HEUER-LEONIDAS SA" on the bridge.
This Camaro features a rare gold-plated bracelet with Heuer signed clasp.

IV. CARRERA PANAMERICANA MEXICO
PORSCHE
DOUBLE VICTORY
in the
1600 c.c. Sports Car Class
1. Herrarte, Guatemala
2. Segura, Argentina
of 14 cars of the 1600 c. c. class
only 2 at the finish: 2 Porsche cars!
DE
CAMINOS
PORSCHE

CARRERA

Bereits 1962 während der Einführung der Autavia kam Jack Heuer die Idee, einen neuen Chronographen für Rennfahrer und Rennsportbegeisterte zu entwickeln. Wichtig war eine gute Ablesbarkeit, ein wasserdichtes Gehäuse, Stoßsicherheit sowie Antimagnetismus. Die Arbeit wurde 1962 begonnen und 1963 der Öffentlichkeit als „Heuer Carrera" präsentiert.
Die Uhr glänzte durch schlichte Eleganz, das Zifferblatt mit zwei oder drei Hilfszifferblättern war extrem klar und leicht ablesbar. Dezimal-, Pulso- und Tachymeter Skalen am Rande des Zifferblattes waren die ersten Varianten einer sehr kreativen Modellpolitik. Wahre Emotion vermittelte der gutklingende Name „Carrera". Aus dem Spanischen kommend, bedeutet das Wort „Wettkampf auf höchstem Niveau".
Jack Heuer wurde zur Namensgebung inspiriert durch die „Carrera Panamericana", das gefährlichste Autorennen der 1950er Jahre. Es führte von der Grenze Mexikos durch Guatemala nach Ciudad Juarez an der nördlichen Grenze zur USA und war eine wahre Materialschlacht über Berge und durch Täler, teils auf unbefestigten, nicht gesperrten Straßen, durch die endlosen Weiten des Mexikanischen Tieflandes und auf den neu errichteten Abschnitten des „Panamerica Highway", den Mexikos Regierung bekannter machen wollte. Enorme Temperaturdifferenzen setzten Mensch und Material sehr zu.
General Motors und Chrysler entsandten ihre Buicks und Plymouths, die dort auf die europäische Konkurrenz von Porsche und Mercedes trafen. Jeder wollte gewinnen, es ging um Prestige und Markenimage, ein Rennen zwischen David und Goliath vor den Toren des wohl wichtigsten amerikanischen Automarktes.
Porsche gibt seither den innovativen, schnellen und auf der Rennpiste erfolgreichsten Modellen mit Recht den Namen „Carrera". Die Verbindung zwischen dem Auto-Motorsport und den Uhren ist bemerkenswert.

During the launch of the Autavia in 1962, Jack Heuer had the idea of creating a new chronograph for drivers and motorsport fans. It would need to be easy to read, shockproof, and anti-magnetic and have a waterproof case. Development started in 1962 and the Heuer Carrera was launched to the public in 1963.
The watch was elegant in its simplicity; the dial with two or three subregisters was extremely clear and easy to read. Decimal, pulso- and tachymeter tracks on the edge of the dial were the first variations in a very creative range of models.
The name "Carrera" breathed the spirit of racing into the brand. The word comes from the Spanish. Carrera means competition at the highest level.
Jack Heuer was inspired to choose this name by the Carrera Panamericana, the most dangerous car race of the 1950s. The Carrera ran from the Mexican border with Guatemala in the south to Ciudad Juarez at the northern border with the USA. It was a terrible battle of man and machine over mountains, unpaved roads, through the wide open spaces of Mexico, and along the newly built sections of the Pan-American Highway that the Government of Mexico sought to promote. Enormous temperature ranges challenged both the drivers and their automobiles.
General Motors and Chrysler sent their Buicks and Plymouths to contend with the European competition from Porsche and Mercedes. All of the participants wanted to win, for the gains in prestige and image a victory would bring. It was a race between David and Goliath on the threshold of the enormous American automobile market.
Since that time, Porsche has named its fastest and most successful innovative racing models "Carrera." A remarkable link between watches and motorsport.
The name "Carrera" was simple, easy to pronounce, and understood at international level; hence, it was

„Carrera“ war einfach, gut aussprechbar und international verständlich und daher ideal für die zukünftigen Wachstumsmärkte. Der Name war stolz wie die Corrida, geprägt von Eleganz und verkörperte den Mythos von Abenteuer, Gefahr und Heldentum. Der große Erfolg der Carrera Modellreihe inspirierte Jack Heuer zu neuen Ideen. 1966 stellte er den ersten Sport-Chronograph mit Datumsanzeige in Form eines bedruckten Rads und Datumsfenster vor.

1968 folgte eine Variante mit Vollkalender in einem dickeren Gehäuse. 1969 gelang Heuer eine Weltsensation: der erste Automatik-Chronograph, angetrieben durch einen Mikro-Rotor, der eine flachere Bauweise erlaubte. Damals eine absolute Neuheit auf dem Uhrenmarkt. Der Gehäusedurchmesser wuchs ganz im Trend der 70er Jahre und konnte bequem das neue Werk aufnehmen, ohne dabei an klassischer Eleganz einzubüßen.

Das Design des Zifferblattes mit zwei kontrastierenden Hilfszifferblättern und der Datumsfunktion bei 6 Uhr wurde von den Käufern sehr gut angenommen.

1970 wurde die Schwingungszahl des Werks von 19.800 auf 21.600 A/h erhöht und von Kaliber 11 auf Kaliber 12 umbenannt. Eine Variante Kaliber 15, mit permanenter kleiner Sekunde, wurde 1972 mit verändertem Zifferblattdesign auf den Markt gebracht. Wieder ganz im Trend der 70er Jahre, gab es ab 1974 eine vergrößerte Gehäuse-Variante mit einem wunderschönen Zifferblatt im Dreibahnenschliff. Die sehr seltene Automatik Carrera aus 18 Karat Gold ist übrigens die Lieblingsuhr von Jack Heuer, die er mit persönlicher Widmung an Formel 1 Fahrer verschenkte.

ideal for future growth markets. The name is as proud as the Corrida, while also exceptionally elegant and evoking the mythical imagery of adventure, danger, and heroism.

The huge success of the Carrera model line inspired Jack Heuer's creative spirit to new heights of innovation. In 1966 he launched the first sport chronograph with date function with wheel and date window, followed in 1968 by a full calendar in a thicker case.

In 1969 Heuer achieved a sensational global success by producing the first automatic chronograph, powered by a micro-rotor, which allowed for a slimmer design. This was an absolute sensation on the watch market.

The case grew in diameter in line with the trend in the early 1970s and could easily house the new movement without losing any of its classic Carrera elegance.

The design of the dial with two contrasting subregisters and date function at 6 o'clock was an immediate success with consumers.

In 1970 the vibration frequency was increased from 19,00 to 21,600 A/h, and Calibre 11 became Calibre 12. A variation was Calibre 15, which featured a permanent slim second hand. It has been produced since 1972 with only slight changes in dial design.

The mid-1970s' trend led to a redesigned larger case in 1974, incorporating a beautiful dial with special grinding finish. The extremely rare 18-karat-gold Carrera Automatic is the preferred watch of Jack Heuer, which he presented to Formula 1 drivers with a personal dedication.

PORSCHE
CARRERA GTS Typ 904
12
12
48a TARGA FLORIO
Classifica generale: 1. 2. 7. 11. 12. GT–2000 cc: 1. 2. 3. 4. 5.
Int. ADAC 1000 km Rennen Nürburgring
Gesamtklassement: 3. 6. 8. 9. 10. 11. 12. 17. 30. GT–2000 ccm: 1. 2. 3. 4. 5. 6. 7. 8. 9.
Les 24 Heures du Mans
Classement général: 7. 8. 10. 11. 12. GT–2000 cc: 1. 2. 3. 4. 5.
12 Heures Internationales de Reims
Classement général: 5. 6. 7. 10. 12. 13. 15. 16. GT–2000 cc: 1. 2. 3. 4. 5. 6. 7. 8.
BP, Dunlop, Bosch
Printed in Germany · Juli 1964

CARRERA

ref. 2447 T „Eggshell“

Erstes Carrera-Gehäuse mit Handaufzug und verschraubtem Boden. Eierschalenfarbiges Zifferblatt mit roter Tachymeter-Skala. Die Hilfszifferblätter für die Minuten und Stunden befinden sich bei 3 bzw. 6 Uhr, die permanente Sekunde bei 9 Uhr. Sie sind in das Zifferblatt eingelassen und weisen konzentrische kreisförmige Rillen auf.

Die Stundenindizes sind aus Stahl mit Facetten auf beiden Seiten und kleinen Leuchtpunkten an den äußeren Enden. Das Carrera-Logo befindet sich über dem Heuer-Schild. Die polierten Stahlzeiger sind mit Leuchtmasse versehen; der Minutenzeiger ist kurz, um die Ablesbarkeit der Tachymeter-Skala zu verbessern. Für die drei Hilfszifferblätter wurden kleine schwarze Zeiger verwendet.

Über 6 Uhr befindet sich der Schriftzug „SWISS“. Das T für Tritium wurde noch nicht auf dem Stundenzifferblatt angebracht, was auf eine Carrera der ersten Serie hinweist.

Die 1/5-Sekunden-Skala ist in den Spannring des Glases integriert.

Die Krone war bei den ersten Carreras unsigniert. Die Chronographen-Drücker befinden sich bei 2 bzw. 4 Uhr. Der charakteristische polygonale Boden und eine größere eingravierte Seriennummer zwischen den Bandanstößen sind weitere Merkmale der frühen Carreras.

Angetrieben wird sie von einem Valjoux 72-Werk mit der Signatur „72 R“ unter der Unruh.

Da Heuer 1964 mit Leonidas fusionierte, sind Übergangssignaturen keine Seltenheit.

First manual-wind Carrera case with screw-back construction. Eggshell dial with red tachymeter scale. The minute and hour subdials are positioned at 3 and 6 o'clock respectively, with a permanent second at 9 o'clock. The subdials are recessed in the dial and have concentric circular ridges.

The hour markers are steel with facets on both sides and small luminous dots at the outer ends. The Carrera logo is positioned over the Heuer shield. The polished steel hands incorporate luminous materials; the minute hand is short to increase the legibility of the tachymeter scale. A black chronograph hand and small black hands were used for the three subdials.

A, "SWISS" signature can be found above 6 o'clock. "T" for tritium was not yet marked on the hour subdial, which indicates a very early Carrera.

The 1/5-second markings are incorporated into the crystal tension ring.

The crown was unsigned on the first run of Carreras. The chronograph pushers are positioned at 2 and 4 o'clock respectively. The characteristic polygonal back and a larger engraved serial number between the lugs are further features of early Carreras.

Powered by a Valjoux 72 movement signed "72 R" under the balance wheel.

Heuer merged with Leonidas in 1964; hence, transitional signatures are not uncommon.

MONTE-CARLO
HEUER
CARRERA
HEUER
SWISS
SEBRING
HEUER

CARRERA

ref. 2447 D

Gehäuse mit verschraubtem Boden. Elfenbeinfarbenes Zifferblatt mit einer blauen Dezimalskala. 30-Minuten- und 12-Stunden-Hilfszifferblätter bei 3 bzw. 6 Uhr, das permanente Sekunden-Hilfszifferblatt bei 9 Uhr. Sie sind versenkt und haben konzentrische Rillen.
Die Stundenindizes sind aus Metall mit Facettenschliff auf beiden Seiten und kleinen Leuchtpunkten am äußeren Ende. Der „CARRERA“-Schriftzug steht über dem Heuer-Logo. Die polierten Stahlzeiger haben mit Leuchtmasse gefüllte Flächen. Der Minutenzeiger ist kurz, damit man die Dezimalskala am Rand des Zifferblattes besser ablesen kann. Der Chronographen-Sekundenzeiger und die kleinen Zeiger der drei Hilfszifferblätter sind schwarz. Am äußeren geneigten Glasfixierungsring ist eine 1/5-Sekunden-Skala aufgedruckt. Die Krone ist mit dem „HEUER“-Logo signiert. Die Chronographen-Drücker sind teilweise zum Schutz ins Gehäuse integriert.
Das Uhrwerk ist ein frühes Kaliber Valjoux 72, signiert mit „72 R“ unter der Unruh und mit „HEUER-LEONIDAS SA“ auf der Brücke.

Case with screw-back construction. Ivory dial and subdials with spectacular blue decimal scale. The 30-minute and 12-hour totalizers are positioned at 3 and 6 o'clock. The permanent second subdial is at 9 o'clock. The subdials are recessed in the dial and have concentric circular ridges.
Smaller hour markers are in metal with facets on both sides and small luminous dots at the outer facet end. The "CARRERA" logo is positioned above the Heuer shield. Polished steel hands have integrated luminous material; the minute hand is short to increase the legibility of the blue decimal scale printed on the edge of the dial. A black-painted chronograph second hand and small black hands were used for the three subdials. The 1/5-second markings are incorporated into the crystal tension ring. The crown is signed "HEUER." The chronograph pushers are partly integrated into the case for protection.
Powered by a Valjoux 72 movement signed "72 R" under the balance wheel and "HEUER-LEONIDAS SA" on the bridge.

CARRERA
HEUER
SWISS
PORSCHE

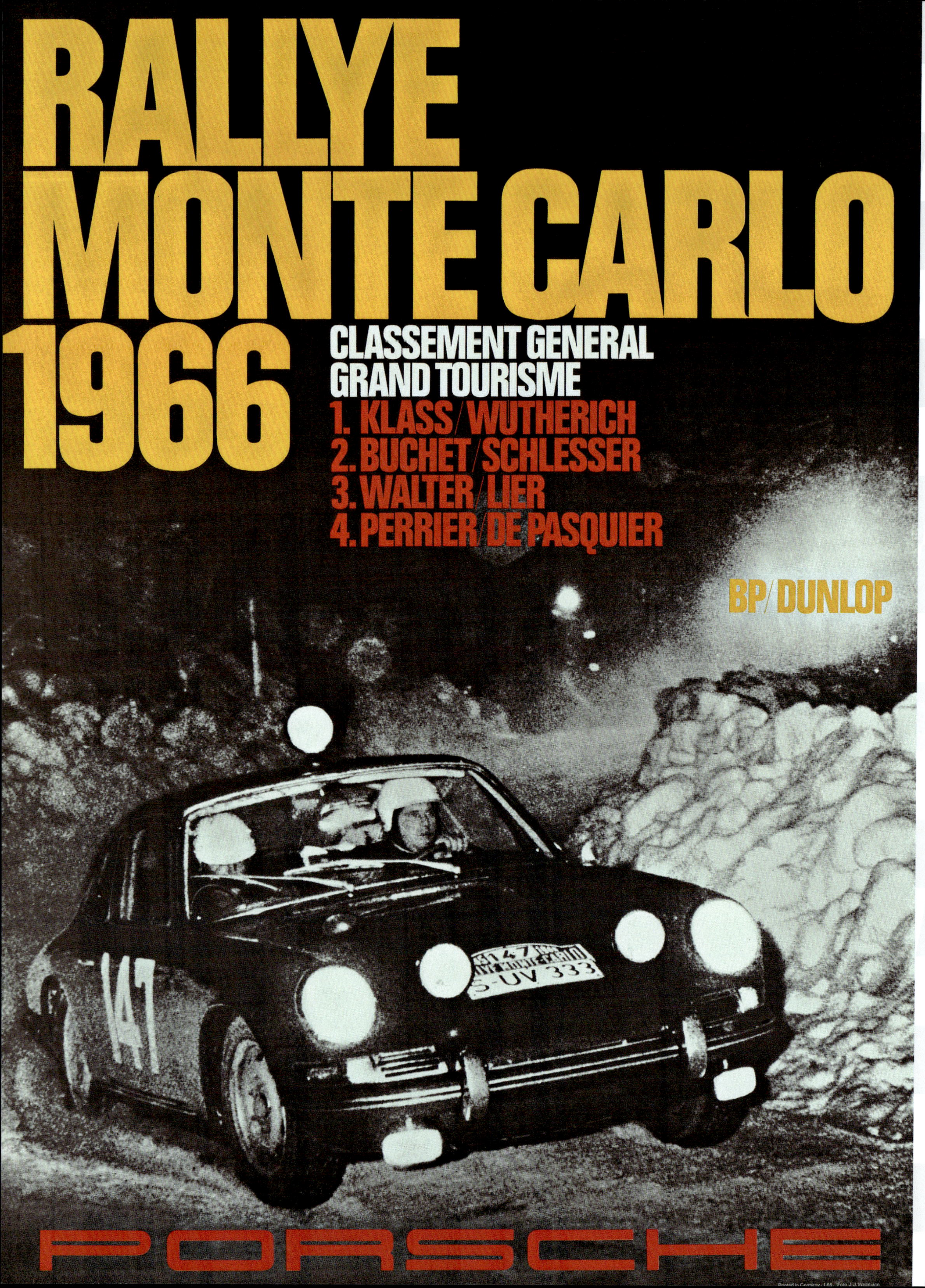
RALLYE
MONTE CARLO
1966
CLASSEMENT GENERAL
GRAND TOURISME
1. KLASS/WUTHERICH
2. BUCHET/SCHLESSER
3. WALTER/LIER
4. PERRIER/DE PASQUIER
BP/DUNLOP
147
S-UV 333
PORSCHE

Professor Ferdinand Alexander Porsche mit seinem Designmodell des 911 Targa.
Professor Ferdinand Alexander Porsche with his 911 Targa design model.

CARRERA

ref. 2447 S „Yachting“

Carrera-Gehäuse mit verschraubtem Boden. Weißes Zifferblatt mit Starburst-Finish und spektakulärem 30-Minuten-Hilfszifferblatt „Yachting“ mit drei blau lackierten Segmenten, die jeweils 5 Minuten anzeigen. Die zwei anderen Zähler für Sekunde und 12 Stunden sind bei 9 bzw. 6 Uhr. Die Hilfszifferblätter sind in das Zifferblatt eingelassen und weisen konzentrische kreisförmige Rillen auf.
Die Stundenindizes sind aus Metall mit Facetten auf jeder Seite und kleinen Leuchtpunkten am Ende. „CARRERA“ steht über dem Heuer-Logo. Polierte Stabzeiger aus Stahl mit integrierter Leuchtmasse. Schwarzer Chronographen-Zeiger und kleine schwarze Zeiger auf den drei Hilfszifferblättern. Die 1/5-Sekundenmarkierungen sind in den Spannring des Glases eingearbeitet. Die Chronographen-Drücker befinden sich bei 2 bzw. 4 Uhr und umgeben die mit Heuer signierte Krone. Der innere Gehäuseboden ist mit „HEUER-LEONIDAS SA“ signiert.
Angetrieben von einem Valjoux 72-Uhrwerk mit der Signatur „72 R“ unter der Unruh und „HEUER LEONIDAS SA“ auf der Brücke.

Carrera case with screw-back. White dial with starburst finish and stunning "Yachting" 30-minute subdial with three blue-painted subsections indicating 5 minutes each. The permanent second subdial is at 9 o'clock and the 12-hour counter is at 6 o'clock. The subdials are recessed in the dial and have concentric circular ridges.
The hour markers are metal with facets on each side and small luminous dots at the end. The "CARRERA" logo is placed above the Heuer shield. The polished baton steel hands have integrated luminous material. Black chronograph hand plus small black hands on the three subdials. The 1/5-second markings are incorporated into the crystal tension ring, The chronograph pushers are positioned at 2 and 4 o'clock surrounding the Heuer signed crown. The inner case back is signed "HEUER-LEONIDAS SA."
Powered by a Valjoux 72 movement signed "72 R" under the balance wheel and "HEUER-LEONIDAS SA" on the bridge.

CARRERA
HEUER

CARRERA

ref. 2456 S

Carrera-Gehäuse aus 18 Karat Gold mit verschraubtem Boden. Weißes Zifferblatt mit Starbust-Finish. Die Hilfszifferblätter für Minuten und Stunden befinden sich bei 3 bzw. 6 Uhr, die permanente Sekunde bei 9 Uhr. Die Hilfszifferblätter sind in das Zifferblatt eingelassen und weisen konzentrische kreisförmige Rillen auf.
Die goldenen Stundenindizes haben Facetten auf jeder Seite und kleine Leuchtpunkte am Ende. Der Modellname „CARRERA" ist über dem Heuer-Logo positioniert. Die goldenen Stabzeiger sind poliert und mit Leuchtmasse gefüllt. Der Chronographen-Sekundenzeiger und die kleinen Zeiger auf den drei Hilfszifferblättern sind goldfarben. Die 1/5-Sekundenmarkierungen sind in den Spannring des Glases eingearbeitet. Die Chronographen-Drücker befinden sich bei 2 bzw. 4 Uhr und umgeben die mit „HEUER" signierte Krone. Der innere Gehäuseboden ist mit „HEUER-LEONIDAS SA" signiert und mit einer Goldpunze versehen.
Angetrieben von einem Valjoux 72-Uhrwerk mit der Signatur „72 R" unter der Unruh und „HEUER LEONIDAS SA" auf der Brücke.

18-karat gold Carrera case with screw back. White dial with a starburst finish. The minute and hour subdials are positioned at 3 and 6 o'clock respectively with a permanent second at 9 o'clock. The subdials are recessed in the dial and have concentric circular ridges.
The gold hour markers have facets on each side and small luminous dots at the end. The "CARRERA" logo is placed above the Heuer shield. Gold baton steel hands are polished with integrated luminous material. The chronograph second hand and the small hands used on the three subdials are golden. The 1/5-second markings are incorporated into the crystal tension ring, The chronograph pushers are positioned at 2 and 4 o'clock surrounding the Heuer signed crown. The inner case back is signed "HEUER-LEONIDAS SA" with a gold hallmark.
Powered by a Valjoux 72 movement signed "72 R" under the balance wheel "72 R" and "HEUER-LEONIDAS SA" on the bridge.

CARRERA
HEUER
SWISS

CARRERA
HEUER
SWISS

CARRERA

ref. 2447 SN

Carrera-Gehäuse mit verschraubtem Boden und Handaufzug. Seltenes silbernes Zifferblatt mit großen schwarzen Hilfszifferblättern. Die 30-Minuten- und 12-Stunden-Totalisatoren befinden sich bei 3 und 6 Uhr, die permanente Sekunde bei 9 Uhr. Das Hilfszifferblatt der permanenten Sekunde befindet sich bei 9 Uhr. Die Hilfszifferblätter sind in das Zifferblatt eingelassen und weisen konzentrische kreisförmige Rillen auf.
Die Stundenindizes aus Metall haben ein neues, konisches Design mit schwarzen Einsätzen und kleinen Leuchtpunkten am äußeren Ende. Das Modell „CARRERA“ ist über dem Heuer-Logo positioniert. Die Zeiger aus poliertem Stahl haben eine integrierte Leuchtmasse und schwarze Einsätze. Ein schwarzer Chronographen-Zeiger und kleine weiße Zeiger wurden für die drei Hilfszifferblätter verwendet. Die 1/5-Sekundenmarkierungen sind in den Spannring des Glases eingearbeitet. Die Chronographen-Drücker sind bei 2 bzw. 4 Uhr positioniert und umgeben die von Heuer signierte Krone. Der innere Gehäuseboden ist mit „HEUER-LEONIDAS SA“ signiert. Angetrieben von einem Valjoux-72-Werk mit der Signatur „72 R“ unter der Unruh und „HEUER LEONIDAS SA“ auf der Brücke.

Carrera case with screw back. Rare silver dial with large black subdials. 30-minute and 12-hour totalizers are positioned at 3 and 6 o'clock. The permanent second subdial is at 9 o'clock. The subdials are recessed in the dial and have concentric circular ridges.
The metal hour markers have a new tapered design, with black inserts and small luminous dots at the outer end. The "CARRERA" logo is positioned above the Heuer shield. The polished steel hands have integrated luminous material and black inserts. A black chronograph hand plus small white hands were used for the three subdials. The 1/5-second-markings are incorporated into the crystal tension ring. The chronograph pushers are positioned at 2 and 4 o'clock surrounding the Heuer signed crown. Inner back case is signed "HEUER-LEONIDAS SA."
Powered by a Valjoux 72 movement signed "72 R" under the balance wheel "72 R" and "HEUER LEONIDAS SA ON THE BRIDGE."

CARRERA
HEUER
SWISS

CARRERA

ref. 2447 NST „James Hunt“

Carrera-Gehäuse mit verschraubtem Boden und Handaufzug. Spektakuläres schwarzes Zifferblatt mit kontrastierenden weißen Hilfszifferblättern und Tachymeter-Skala. Die 30-Minuten- und 12-Stunden-Totalisatoren befinden sich bei 3 und 6 Uhr, die permanente Sekunde bei 9 Uhr. Die Hilfszifferblätter sind in das Zifferblatt eingelassen und weisen konzentrische kreisförmige Rillen auf.
Die Stundenindizes aus Metall haben ein neues, konisches Design mit weißen Einsätzen und kleinen Leuchtpunkten am äußeren Ende. Der Modellname „CARRERA“ ist über dem Heuer-Logo positioniert. Die Stabzeiger aus poliertem Stahl haben eine integrierte Leuchtmasse und schwarze Einsätze. Es wurden ein weißer Chronographen-Sekundenzeiger und drei kleine schwarze Zeiger für die Hilfszifferblätter verwendet.
Die 1/5-Sekundenmarkierungen sind in den Spannring des Glases eingearbeitet. Die Chronographen-Drücker befinden sich bei 2 bzw. 4 Uhr und umgeben die Krone mit Heuer-Signatur. Der innere Gehäuseboden ist mit „HEUER-LEONIDAS SA“ signiert.
Angetrieben von einem Valjoux 72-Uhrwerk mit der Signatur „72 R“ unter der Unruh und „HEUER LEONIDAS SA“ auf der Brücke. Ein ähnliches Modell mit Guy Freres-Stahlband (wie hier abgebildet) sieht man auf historischen Fotos und in einer Vauxhall-Werbung an Formel 1-Champion James Hunt.

Manual-wind Carrera case with screw back. Stunning black dial with contrastive white subdials and tachymeter scale. The 30-minute and 12-hour totalizers are positioned at 3 and 6 o'clock. The permanent second subdial is at 9 o'clock. The subdials are recessed in the dial and have concentric circular ridges.
The metal hour markers have a new tapered design, with black inserts and small luminous dots at the outer end. The "CARRERA" name is positioned above the Heuer shield. The polished steel baton hands have integrated luminous material and black inserts. A white chronograph second hand plus three small black hands were used for the subdials. The 1/5-second markings are incorporated into the crystal tension ring. The chronograph pushers are positioned at 2 and 4 o'clock surrounding the Heuer signed crown. The inner case back is signed "HEUER-LEONIDAS SA."
Powered by a Valjoux 72 movement signed "72 R" under the balance wheel and "HEUER-LEONIDAS SA" on the bridge. A similar watch on a Guy Freres steel band (Shown here) was worn by Formula 1 Champion James Hunt in period photographs and a Vauxhall advertising shoot.

Volvo war in den 70er Jahren sehr aktiv im Motorsport. Eine Autavia und zwei Carreras, bedruckt mit dem Volvo-Logo auf dem Zifferblatt.

Volvo was very active in motorsport throughout the 70s. Here are an Autavia and two Carreras featuring the Volvo logo on the dials.

CARRERA

ref. 3647 N „MG“

Gehäuse mit verschraubtem Boden. Matt-schwarzes Zifferblatt mit weißem MG-Logo und speziell angefertigtem Tachymeterring. Das permanente Sekunden- und 45-Minuten-Hilfszifferblatt bei 3 bzw. 9 Uhr. Die Hilfszifferblätter sind vertieft und haben konzentrische Rillen.
Die metallenen Stundenindizes haben Facettenschliff auf beiden Seiten, mit kleinen Leuchtpunkten am äußeren Ende. Doppelt breite Indizes bei 12 Uhr und kleine quadratische bei 3 und 9 Uhr, um den Hilfszifferblättern Raum zu geben. Der „CARRERA“-Schriftzug in Weiß steht über dem Heuer-Logo. Das „MG“-Logo und der Name „E. W. GONZALEZ“ ist über 6 Uhr gedruckt. Der lange polierte Stahl-Minutenzeiger reicht bis ans äußere Ende der Stundenindizes. Der Chronographen-Sekundenzeiger und die kleinen Zeiger der zwei Hilfszifferblätter sind weiß. Die Krone ist mit dem Heuer-Logo signiert. Die Chronographen-Drücker sind vergrößert und speziell adaptiert worden, um das Stoppen mit Handschuhen zu erleichtern. Der verschraubte Boden der Carrera ist auf der Innenseite mit „HEUER-LEONIDAS SA“ signiert.
Das Uhrwerk ist ein Valjoux Kaliber 92, signiert mit „92 R“ unter der Unruh und mit „HEUER-LEONIDAS SA“ auf der Brücke.

Case with screw-back construction. Matt black dial with printed white MG logo and custom tachymeter ring. The permanent second and 45-minute totalizers are positioned at 3 and 9 o'clock. The subdials are recessed in the dial and have concentric circular ridges.
Metal hour markers are bevelled on both ends with small luminous dots on the outer bevel. Parallel indices are positioned at 12 o'clock and small square ones at 3 and 9 o'clock to allow space for the subdials. "CARRERA" is printed in white over the Heuer logo. The "MG" logo and the name "E. W. GONZALEZ" are printed above 6 o'clock. The long polished steel minute hand reaches the outer bevel of the hour markers. A white-painted chronograph second hand and small white hands were used for the two subdials. The crown is signed with the HEUER logo. The chronograph pushers are enlarged to simplify timing with gloves. This Carrera has a screw back; the interior of the case is signed "HEUER-LEONIDAS SA."
Powered by a Valjoux 92 movement signed "92 R" under the balance wheel and "HEUER-LEONIDAS SA" on the bridge.

TACHYMETRE
CARRERA
HEUER
MG
T SWISS
E.W. GONZALEZ

CARRERA

ref. 3147 S „Indianapolis“

Gehäuse mit verschraubtem Boden. Das Carrera-Modell mit Datum trägt das „INDIANAPOLIS MOTOR SPEEDWAY“-Logo auf dem Zifferblatt mit Sternschliff, weißen Hilfszifferblättern und einer Datumsanzeige. Das permanente Sekunden- und das 45-Minuten-Hilfszifferblatt sind bei 3 bzw. 9 Uhr. Die Hilfszifferblätter sind vertieft und haben konzentrische Rillen.
Die Stundenindizes sind aus Metall, mit Facettenschliff auf beiden Seiten und kleinen Leuchtpunkten am äußeren Ende. Zwei parallele Indizes sind bei 12 Uhr. Das Datumsfenster mit roten Ziffern steht unter der 12. „CARRERA“- und Heuer-Logo sind gegen die Mitte hin verschoben, um dem Datumsfenster Raum zu geben. Das „INDIANAPOLIS MOTOR SPEEDWAY“-Logo ist über 6 Uhr gedruckt und entspricht in der Originalfarbe dem registrierten Markenzeichen ® der berühmten amerikanischen Rennstrecke in Indiana. Die polierten Stahlzeiger haben mit Leuchtmasse gefüllte Flächen. Der Chronographen-Sekundenzeiger und die kleinen Zeiger der zwei Hilfszifferblätter sind schwarz. Am äußeren geneigten Glasfixierungsring ist eine 1/5-Sekunden-Skala aufgedruckt. Die Krone ist unsigniert.
Das Uhrwerk ist ein Landeron Kaliber 189, signiert mit „L 189“ unter der Unruh und mit „HEUER-LEONIDAS S.A.“ auf der Brücke.

Case with screw back. Carrera Date has the "INDIANAPOLIS MOTOR SPEEDWAY" logo printed on the dial with starburst finish, white subdials, and a date feature. The permanent second and 45-minute totalisers are positioned at 3 and 9 o'clock. The subdials are recessed in the dial and have concentric circular ridges.
The metal hour markers are bevelled on both ends with small luminous dots on the outer bevel. Parallel markers are used at 12 o'clock and small rectangular ones at 3 and 9 o'clock to allow space for the subdials. The date window with red numerals is below the shortened twin facets at 12 o'clock. "CARRERA" and the Heuer logo are slightly closer to the center to allow space for the date window. The "INDIANAPOLIS MOTOR SPEEDWAY" logo is printed above 6 o'clock, with the original logo color codes and ® for the registered trademark of the famous racetrack in Indiana, USA. The polished steel hands have integrated luminous material. A black painted chronograph second hand and small black hands were used for the two subdials. The 1/5-second markings are incorporated into the crystal tension ring. The crown is unsigned.
Powered by a Landeron 189 movement signed "L 189" under the balance wheel and "HEUER-LEONIDAS S.A." on the bridge.

8
CARRERA
HEUER
60
40
20
45
30
15
INDIANAPOLIS
MOTOR
SPEEDWAY
T
SWISS

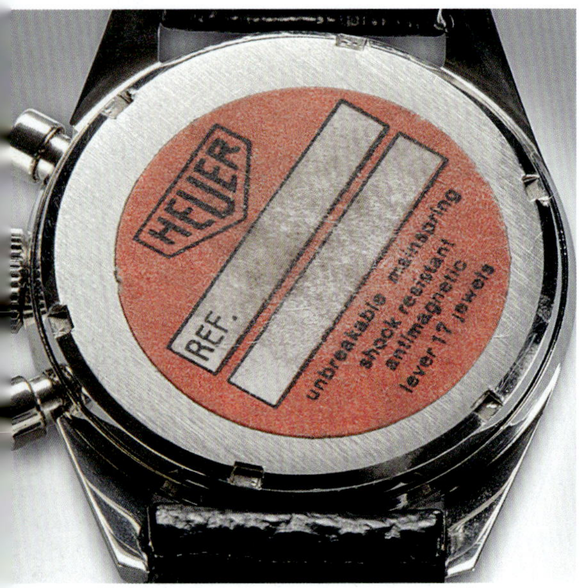

CARRERA

ref. 3147 N „Indianapolis“

Gehäuse mit verschraubtem Boden. Version des Carrera-Datumsmodells mit seltenem „INDIANAPOLIS MOTOR SPEEDWAY“-Logo auf dem schwarzen Zifferblatt, mit kontrastierendem weißem Hilfszifferblatt und Datumsanzeige. Das permanente Sekunden-Hilfszifferblatt fehlt und machte dem Datumsfenster bei 9 Uhr Platz. Das 45-Minuten-Hilfszifferblatt bleibt bei 3 Uhr. Das Hilfszifferblatt ist vertieft und hat konzentrische Rillen.
Die Stundenindizes sind aus Metall, mit Facettenschliff auf beiden Seiten und kleinen Leuchtpunkten am äußeren Ende. Zwei parallele Indizes sind bei 12 Uhr. Die Indizes bei 3 und 9 Uhr fehlen. Das Datumsfenster mit schwarzen Ziffern steht bei 9 Uhr.
Der „CARRERA“-Schriftzug steht über dem Heuer-Logo. Das „INDIANAPOLIS MOTOR SPEEDWAY“-Logo ist über 6 Uhr gedruckt. Die polierten Stahlzeiger haben mit Leuchtmasse gefüllte Flächen. Der Chronographen-Sekundenzeiger ist weiß, der kleine Zeiger des Minuten-Hilfszifferblattes schwarz.
Am äußeren geneigten Glasfixierungsring ist eine 1/5-Sekunden-Skala angebracht. Die Krone ist unsigniert. Der verschraubte Boden hat einen Längsschliff. Die ungetragene Uhr zeigt noch das rote Heuer-Etikett auf dem Boden. Das Uhrwerk ist ein Landeron Kaliber 189, signiert mit „L 189“.

Case with screw back. This version of the Carrera Date features the rare printed "INDIANAPOLIS MOTOR SPEEDWAY" logo, stunning black dial, contrasting white subdial, and a date function. The permanent second subdial was replaced by the date window repositioned at 9 o'clock. The 45-minute totalizer remains at 3 o'clock. The subdial is recessed in the dial and has concentric circular ridges.
Metal hour markers are bevelled on both ends with small luminous dots on the outer bevel. Parallel markers are positioned at 12 o'clock but omitted at 9 and 3 o'clock to provide space for the date window with black numerals and the subdial. The "CARRERA" name is printed above the Heuer logo below 12 o'clock.
The "INDIANAPOLIS MOTOR SPEEDWAY" logo is printed above 6 o'clock. The polished steel hands have integrated luminous material. A white-painted chronograph second hand and a small black hand were used on the minute recording subdial.
The 1/5-second markings are incorporated into the crystal tension ring. The crown is unsigned. Powered by a Landeron 189 movement signed "L189."

CARRERA
HEUER
45
30
15
8
INDIANAPOLIS
MOTOR
SPEEDWAY
SWISS

CARRERA

ref. 2547 N „Gerd-Rüdiger Lang“

Die Carrera 12 Dato ist der umfangreichste und komplizierteste Chronograph der Modellreihe. Sie kombiniert einen Vollkalender, der Datum, Tag und Monat anzeigt, mit einer 1/5-Sekunden-Stoppuhr, deren kontrastierende Hilfszifferblätter bis zu 12 Stunden stoppen und anzeigen können.

Das Gehäuse ist dicker als das der Standard-Carrera, mit anderen Bandanstößen und kleinen Stiftlöchern zum Einstellen der Kalenderfunktionen auf der linken Seite bei 8 und 10 Uhr.

Das Hilfszifferblatt der permanenten Sekunde befindet sich bei 9 Uhr. Auf dem schwarzen Zifferblatt mit verkürzten Indizes aus Stahl ist die Zahl 12 als Ziffer geschrieben. Darunter befinden sich die Fenster für Tag und Monat, darunter das Heuer-Logo.

Die Modellbezeichnung Carrera wurde aus Platzgründen weggelassen.

Die Hilfszifferblätter sind an den Stellen abgeschrägt, an denen die Sekunden-, Minuten- und Stundenindizes aufgedruckt sind, und weisen konzentrische kreisförmige Rillen auf. Ein leicht vertiefter weißer Randstreifen am Zifferblatt zeigt eine schwarze 1/5-Sekunden-Skala, sowie das Datum mit blauen Ziffern vom 1. bis zum 31.

Die polierten Stahlzeiger sind mit integrierter Leuchtmasse versehen; das Datum wird durch einen weiß lackierten Datumszeiger mit rotem Halbmond angezeigt. Es wurden ein maßgeschneiderter (Rolex Daytona) Chronographen-Sekundenzeiger aus Stahl und kleine schwarze Zeiger auf den drei Hilfszifferblättern verwendet. Die Drücker sind asymmetrisch, was dazu führt, dass der Startdrücker näher an der Krone liegt als der Nullstellungsdrücker. Die Krone ist mit „HEUER“ signiert.

Die Carrera 12 Dato hat einen flachen Gehäuseboden mit gebürsteter Oberfläche.

Angetrieben wird sie von einem Valjoux 723-Uhrwerk mit den Signaturen „723 R“ unter der Unruh, und „HEUER LEONIDAS SA“ auf der Brücke. Diese Uhr war Eigentum des unabhängigen Uhrmachers Gerd-Rüdiger Lang. Die neue Seriennummer 119911 hatte er aus Spaß und als Hommage an einen Porsche 911 angebracht.

The Carrera 12 Dato is the most comprehensive and complicated chronograph of the model line. It combines a full calendar displaying date, day, and month with a 1/5-second chronograph recording up to 12 hours on contrasting subdials. The case is thicker than the standard Carrera case and different in shape. As a unique design feature, there are small pin holes at 8 and 10 o'clock on the left for setting the date functions.

The permanent second subdial is at 9 o'clock. The black dial with shortened steel hour markers shows the number 12, with the day and month windows and Heuer logo below.

The Carrera model name was omitted due to lack of space.

The subdials are sloped at the second, minute, and hour markers and have concentric circular ridges. A slightly recessed white outer circle on the dial indicates the black 1/5-second track. The date is printed in blue digits from the 1st to the 31st.

The polished steel hands have integrated luminous material; the date is shown by the red crescent moon at the end of the white-painted date hand. A customized (Rolex Daytona) steel chronograph second hand and small black hands were used on the three subdials. The pushers are asymmetric, so that the starting pusher is closer to the crown than the reset pusher. The crown is signed "HEUER."

The Carrera 12 Dato has a flat case back with a brushed finish.

Powered by a Valjoux 723 movement signed "723 R" under the balance wheel and "HEUER LEONIDAS SA" on the bridge. It was formerly owned by independent watchmaker Gerd-Rüdiger, who applied a new seriel number 119911 as a tribute to the Porsche 911.

SAT
DEC
HEUER
SWISS
MADE

CARRERA

ref. 73653 N „Pulsometer“

Zweite Generation des Carrera-Gehäuses mit verschraubtem Gehäuseboden und Handaufzug, vergrößert auf 38,5 mm. Graublaues Zifferblatt mit weißen Hilfszifferblättern, weiß gedruckter 1/5-Sekunden-Skala am äußeren Rand und einem abgeschrägten Spannring mit einer spektakulären blauen Pulsometer-Skala.
Die 30-Minuten- und 12-Stunden-Totalisatoren befinden sich bei 3 und bei 6 Uhr, die permanente Sekunde ist bei 9 Uhr positioniert. Die Hilfszifferblätter sind leicht in das Zifferblatt eingelassen und zeigen feine konzentrische kreisförmige Rillen.
Abgeschrägte Stundenindizes aus Stahl mit weißen Einsätzen und einem Leuchtstoffbalken an der Außenseite. Das Carrera-Logo ist weiß und über dem Heuer-Logo unterhalb von 12 Uhr positioniert. Die Stahlzeiger haben schwarze Einsätze und eine integrierte Leuchtmasse. Es wurden ein roter Chronographen-Zeiger sowie auf den kontrastierenden weißen Hilfszifferblättern drei kleine schwarze Zeiger verwendet.
Die Krone ist mit einem Heuer-Logo versehen, und die Chronographen-Drücker sind tief in das Gehäuse integriert, aber auf der Rückseite gut sichtbar.
Das Uhrwerk ist ein Valjoux 7736-Werk mit der Signatur „7736 R“ unter der Unruh und „HEUER LEONIDAS SA“ auf der Brücke.

Second-generation manual wind Carrera screw-back case enlarged to 38.5 mm (1.52 in). The gray-blue dial has white subdials, white printed 1/5-second scale at the outer edge, and an angled tension ring featuring a striking blue printed pulsometer scale.
The 30-minute and 12-hour totalisers are positioned at 3 and 6 o'clock. The permanent second subdial is at 9 o'clock. The subdials are slightly recessed in the dial revealing fine concentric circular ridges.
Beveled steel hour markers with white inserts and luminous bars at the outer end. The Carrera name is printed in white and positioned above the Heuer logo below 12 o'clock. The baton steel hands have black inserts plus integrated luminous material. A red chronograph second hand and three small black hands were used on the contrasting white subdials.
The crown is signed "HEUER." The chronograph pushers are deeply integrated into the case but fully visible from the back.
Powered by a Valjoux 7736 movement signed "7736 R" under the balance wheel and "HEUER LEONIDAS SA" on the bridge.

BASE 30 PULSATIONS
CARRERA
HEUER
SWISS

CARRERA

ref. 73453 S

Zweite Generation des Carrera-Gehäuses mit verschraubtem Boden und Handaufzug, vergrößert auf 38,5 mm. Silberfarbenes Zifferblatt mit zwei schwarzen Hilfszifferblättern, einer schwarz gedruckten 1/5-Sekunden-Skala am äußeren Rand und einem schwarzen Spannring mit weiß gedruckter Tachymeter-Skala.
Die permanente Sekunde und das 30-Minuten-Zifferblatt befinden sich bei 9 und bei 3 Uhr. Die Hilfszifferblätter sind leicht in das Zifferblatt eingelassen und weisen feine konzentrische kreisförmige Rillen auf.
Abgeschrägte Stundenindizes aus Stahl mit schwarzen Einsätzen und einem Leuchtstoffbalken an der Außenseite. Das Carrera-Logo ist schwarz und über dem Heuer-Logo unterhalb von 12 Uhr positioniert. Die Zeiger aus gebürstetem Stahl sind an den Seiten und Spitzen schwarz und haben eine integrierte Leuchtmasse. Es wurden ein orangefarbener Chronographen-Zeiger und zwei kleine weiße Zeiger für die kontrastierenden schwarzen Hilfszifferblätter verwendet. Bei 6 Uhr befindet sich ein silbern umrahmtes Datumsfenster.
Die Krone ist mit dem Heuer-Logo versehen, und die Chronographen-Drücker sind tief in das Gehäuse integriert, aber auf Rückseite gut sichtbar.
Das Uhrwerk ist ein Valjoux 7734 mit der Signatur „7734 R“ unter der Unruh und „HEUER LEONIDAS SA“ auf der Brücke.

Second-generation Carrera screw-back case enlarged to 38.5 mm (1.52 in). silver-plated dial has black subdials, a black printed 1/5-second scale at the outer edge, and a black tension ring featuring a white printed tachymeter scale.
The permanent second and 30-minute subdials are positioned at 9 and 3 o'clock. The subdials are slightly recessed in the dial and have fine concentric circular ridges.
The beveled steel hour markers have black inserts and a luminous bar at the outer side. The Carrera name is printed in black and positioned above the Heuer logo below 12 o'clock. The brushed steel hands have black print at sides and tips with integrated luminous inlay. An orange chronograph second hand and two small white hands were used on the contrasting black subdials. There is a silver-framed date window at 6 o'clock.
The crown is signed "HEUER." The chronograph pushers are deeply integrated into the case but fully visible from the back.
Powered by a Valjoux 7734 movement signed "7734 R" under the balance wheel and "HEUER LEONIDAS SA" on the bridge.

TACHY
CARRERA
HEUER
SWISS

CHRONOMATIC
HEUER
CARRERA
SWISS
TACHY

CARRERA

ref. 1153 N

Erstes Automatik-Gehäuse mit verschraubtem Boden und Strichschliff auf der Gehäuseoberseite und an den Flanken. Das Zifferblatt ist dunkelblau mit weißen Hilfszifferblättern, die 30 Minuten bzw. 12 Stunden anzeigen, und von einem weißen Tachymeter-Ring mit schwarzen Zahlen umgeben.

Der Schriftzug „CHRONOMATIC" und das Heuer-Logo stehen unter 12 Uhr. „CARRERA" steht über dem weiß gerahmten Datumsfenster. Indizes aus Metall mit weißen Einlagen sowie einem Leuchtbalken am äußeren Ende. Zwei parallele Indizes bei 12 Uhr, die bei 3, 6 und 9 Uhr fehlen.

Das 12-Stunden-Hilfszifferblatt zeigt die Zahlen 3, 6, 9 und 12 einer frühen Version, die spätere Version zeigt die Zahlen durchgehend von 1 bis 12. Die Hilfszifferblätter sind bei den Minuten- und Stunden-Einteilungen abgeschrägt, die Planfläche hat konzentrische Rillen.

Die polierten Stahlzeiger haben mit Leuchtmasse gefüllte Flächen und schwarze Einlagen zur Mitte hin. Der rote Chronographen-Sekundenzeiger reicht präzise zu der 1/5-Sekunden-Skala am äußeren Rand des Zifferblattes.

Die Krone ist links und mit „HEUER" signiert. Die Drücker sind rechts teilweise im Gehäuse integriert. Der Rückendeckel zeigt einen kreisförmigen Schliff. Die Uhr, angetrieben von einem Kaliber 11 Automatik-Uhrwerk, ist mit „HEUER CAL. 11" auf der Brücke signiert.

First Automatic screw-back case design, with brushed finish on top and sides. The dial is deep blue with white subdials recording up to 30 minutes and 12 hours, and is surrounded by a white tachymeter ring with black numerals.

"CHRONOMATIC" and the Heuer logo are below 12 o'clock. "CARRERA" is printed above the white-framed date window.

The metal hour markers have white inserts and luminous bars at the outer end. Parallel markers are positioned at 12 o'clock and markers are omitted at 3, 6, and 9 o'clock. The hour register shows numerals 3, 6, 9, and 12, indicating an early dial version; the later version has all numerals from 1 to 12.

The subdials are beveled at the minute and hour divisions; their faces are flat with concentric circular ridges. The polished steel hands have luminous areas and fine black inserts toward the center. The red chronograph hand reaches the 1/5-second track precisely on the outer edge of the dial.

The crown is on the left and signed "HEUER." The pushers are integrated on the right of the case. The case back has a circular grinding finish.

The watch is powered by a Calibre 11 automatic movement and is signed "HEUER CAL. 11" on the bridge.

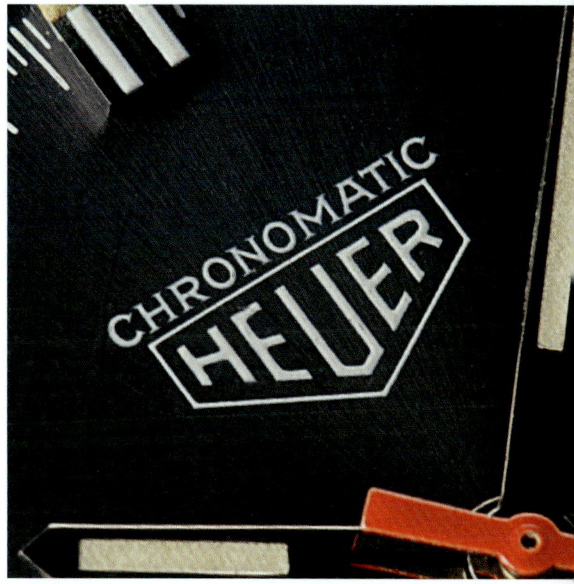

CARRERA

ref. 1153 S

Ein Carrera-Automatikmodell der frühen 70er Jahre mit verschraubtem Gehäuse und gebürsteter Oberfläche auf der Oberseite und den Seiten. Das „Sunburst“-Zifferblatt ist silberfarben mit zwei schwarzen Hilfszifferblättern, die 30 Minuten und 12 Stunden anzeigen. Es wird von einem schwarzen Tachymeter-Ring mit weißen Ziffern umrahmt. Unterhalb von 12 Uhr wurde „CARRERA“ über dem Heuer-Logo neu positioniert, „CHRONOMATIC“ verschwand und wurde fortan von Breitling verwendet. Heuer verwendet von nun an „AUTOMATIC CHRONOGRAPH“ (von Jack Heuer als selbsterklärend angesehen) über dem Datumsfenster bei 6 Uhr. Die Stundenindizes aus Metall haben schwarze Einsätze und leuchtende Balken am äußeren Ende. Zwei parallele Indizes befinden sich bei 12 Uhr, während die Indizes bei 3, 6 und 9 Uhr weggelassen wurden. Das 12-Stunden-Hilfszifferblatt zeigt die Ziffern 3, 6, 9 und 12, was auf die frühe Zifferblattvariante hindeutet, die spätere zeigt alle Ziffern von 1 bis 12.
Die Hilfszifferblätter sind abgeschrägt und haben konzentrische kreisförmige Rillen. Die Stahlzeiger sind mit Leuchtmasse und feinen schwarzen Einsätzen in der Mitte versehen. Der rote Chronographen-Zeiger erreicht die 1/5-Sekunden-Skala genau am äußeren Rand des Zifferblatts. Die Krone befindet sich auf der linken Seite und ist mit „HEUER“ signiert. Die Chronographen-Drücker sind auf der rechten Seite des Gehäuses integriert, aber von der Rückseite aus vollständig sichtbar. Der Gehäuseboden hat einen Rundschliff.
Die Uhr wird von einem Automatikwerk Kaliber 11 angetrieben und ist auf der Brücke mit „HEUER CAL 11“ signiert.

An early 70is Automatic Carrera example with screw-back case and brushed finish on top and sides. The starburst dial is silver with black subdials recording up to 30 minutes and 12 hours. It is encircled by a black tachymeter ring with white numerals. Underneath 12 o'clock, the "CARRERA" name repositioned above the Heuer logo and "CHRONOMATIC" was replaced by "AUTOMATIC CHRONOGRAPH" (seen as more self-explanatory by Jack Heuer) above the date window at 6 o'clock. Henceforth, "CHRONOMATIC" was used by Breitling. The metal hour markers have black inserts and luminous bars at the outer end. Twin parallel markers are positioned at 12 o'clock and markers have been omitted at 3, 6, and 9 o'clock. The hour register exhibits the numbers 3, 6, 9, and 12, indicating an early dial model; the later version shows the complete numerals from 1 to 12.
The subdials are angled with concentric circular ridges. The baton steel hands have luminous and fine black inserts towards the centre. The red chronograph hand reaches the 1/5-second-track precisely on the outer edge of the dial. The crown is on the left and signed "HEUER." The chronograph pushers are integrated on the right of the case but fully visible from the reverse. The case back has a circular grinding finish.
The watch is powered by a Calibre 11 automatic movement and is signed "HEUER CAL 11" on the bridge.

CARRERA
HEUER
AUTOMATIC
CHRONOGRAPH
SWISS
TACHY

1973 TARGA FLORIO WINNER
1. Müller / van Lennep Martini-Porsche Carrera RSR
2. Munari / Andruet Lancia Stratos
3. Kinnunen / Haldi Martini-Porsche Carrera RSR
4. McBoden / Moreschi Chevron
5. Nicodemi / Moser Lola
6. Steckkönig / Pucci Martini-Porsche Carrera RSR
Shell
Bosch / Bilstein / Dunlop
MARTINI INTERNATIONAL CLUB
MARTINI
MARTINI RACING
PORSCHE
STUTTGART
PORSCHE
Dr.-Ing. h.c. F. Porsche A.G. Stuttgart-Zuffenhausen · MARTINI & ROSSI AG Bad Kreuznach · Entwurf Atelier Strenger · Foto Dr. Biesenberger · Printed in Germany · Poster Nr. 2

Jack Heuer präsentiert sein neues Uhrensortiment in Deutschland.

Jack Heuer presenting his new watch portfolio in Germany.

TACHY
UHRMACHERSCHULE
SOLOTHURN
AUTOMATIC
CHRONOGRAPH
SWISS
HEUER

CARRERA

ref. 1153 S „Uhrmacherschule Solothurn“

Automatik-Gehäuse mit verschraubtem Boden, oben und an den Seiten zu Übungszwecken poliert. Das Zifferblatt ist silberfarben mit schwarzen 30-Minuten- und 12-Stunden-Hilfszifferblättern, umgeben von einem anthrazitfarbenen Tachymeter-Ring mit weißen Ziffern.
„UHRMACHERSCHULE SOLOTHURN“ ersetzt den Carrera-Modellnamen und das Heuer-Logo. Über dem Datumsfenster bei 6 Uhr ist „AUTOMATIC CHRONOGRAPH“ aufgedruckt. Parallele Doppel-Stundenindizes aus Metall (zweite Ausführung) haben schwarze Einsätze und Leuchtbalken am äußeren Ende. Bei 3, 6 und 9 Uhr befinden sich keine Indizes.
Das 12-Stunden-Hilfszifferblatt zeigt die Ziffern 1 bis 12, was auf eine spätere Zifferblattversion hinweist. Die Hilfszifferblätter sind abgeschrägt und haben konzentrische kreisförmige Rillen.
Die Zeiger aus gebürstetem Stahl sind leicht vergrößert und an den Außenseiten breiter, mit Leuchtmasse und schwarzer Umrandung. Der orangefarbene Chronographen-Zeiger erreicht die 1/5-Sekunden-Skala am äußeren Rand des Zifferblatts. Die Krone befindet sich auf der linken Seite und ist mit „HEUER“ signiert. Runde Chronographen-Drücker sind an der rechten Seite des Gehäuses integriert, von der Rückseite jedoch vollständig sichtbar. Der Gehäuseboden ist poliert. Die Uhr wird von einem Automatikwerk Kaliber 12 angetrieben.
Dieses seltene Stück wurde von einem Schüler der Solothurner Uhrmacherschule speziell für seine Abschlussprüfung zum Uhrmacher angefertigt.

Automatic Carrera screw-back case with top and sides polished as a practice piece. The dial is silver with black subdials recording 30 minutes and 12 hours, encircled by a charcoal tachymeter ring with white numerals.
"UHRMACHERSCHULE SOLOTHURN" replaced the Carrera model name and Heuer shield logo. "AUTOMATIC CHRONOGRAPH" is printed above the date window at 6 o'clock.
The parallel twin metal hour markers (second model) have black inserts and luminous bars at the outer end. Markers are omitted at 3, 6 and 9 o'clock. The hour register shows the numerals 1 to 12, indicating a later dial version. The subdials are angled with concentric circular ridges. The brushed steel hands are slightly enlarged with wider luminous and black printing towards the outer sides. The orange chronograph hand reaches the 1/5-second-track on the outer edge of the dial. The crown is on the left and signed "HEUER." Round chronograph pushers are integrated on the right of the case but fully visible from the back. The case back has a polished finish. The watch is powered by a Calibre 12 Automatic movement.
This rare piece was especially crafted by an apprentice at the Solothurn watchmaker school as a submission for qualification as a watchmaker.

HEUER
CARRERA
HEUER
TACHY
AUTOMATIC
CHRONOGRAPH
SWISS
18

CARRERA

ref. 1158 CHN „Black Date“

18-Karat-Gehäuse mit verschraubtem Boden und Strichschliff auf der Gehäuseoberseite und an den Flanken. Die Goldpunze (18 Kt.) sitzt zwischen den Chronographen-Drückern auf der rechten Seite. Das Zifferblatt ist goldfarben mit schwarzen Hilfszifferblättern, die 30 Minuten bzw. 12 Stunden anzeigen, und das von einem goldenen Tachymeter-Ring mit schwarzen Zahlen umgeben ist. Bei 12 Uhr ist ein rechteckiger Goldindex mit schwarzem Steineinschluss. Der „CARRERA“-Modellname steht über dem Heuer-Logo und „AUTOMATIC CHRONOGRAPH“ über dem schwarzen Datumsfenster mit weißer Zahl.
Feine goldene Parallel-Indizes haben schwarze Einlagen und Leuchtpunkte am äußeren Ende, keine Indizes bei 3, 6 und 9 Uhr. Jede Zahl des 12-Stunden-Hilfszifferblatts ist in Gold gedruckt. Die Hilfszifferblätter sind abgeschrägt und haben konzentrische Rillen. Gebürstete Goldzeiger haben mit Leuchtmasse gefüllte Flächen, die schwarze Umrahmungen und Spitzen aufweisen. Der schwarze Chronographen-Zeiger reicht präzise zur 1/5-Sekunden-Skala. Die Krone ist mit „HEUER“ signiert. Die Drücker sind im Gehäuse integriert. Der innere Gehäuseboden verfügt über einen Rundschliff und eine Goldpunze.
Die Uhr wird von einem Automatikwerk angetrieben, das auf der Brücke mit Heuer signiert ist.

18-karat gold screw-back case with brushed finish on top and sides. A hallmark is between the chronograph pushers at the right. The dial is gold with black subdials recording 30 minutes and 12 hours and is surrounded by a gold tachymeter ring with black numerals. At 12 o'clock there is a beautiful rectangular gold index with a black stone insert. "CARRERA" is shown over the Heuer logo and "AUTOMATIC CHRONOGRAPH" is printed above the black date window in white lettering.
The fine twin gold hour markers have black inserts and luminous dots at the outer end. Markers have been omitted at 3, 6, and 9 o'clock. The hour register has each individual number painted in gold. The subdials are angled, the face is flat with concentric circular ridges. The brushed gold hands with luminous material have black printing towards the outer sides and black tips. The black chronograph hand reaches the 1/5-second track on the outer edge of the dial. The crown is signed "HEUER." The pushers are integrated. The inner case back has a circular grinding finish and a gold hallmark.
The watch is powered by an Automatic movement signed "HEUER" on the bridge.

Jacky Ickx und Clay Regazzoni neben beschädigtem Ferrari mit Heuer-Logo, 1971.

Jacky Ickx und Clay Regazzoni next to a wrecked Ferrari with Heuer logo in 1971.

Jack Heuer schenkte jedem Ferrari-Piloten eine goldene Carrera, mit im Rückdeckel gravierten Namen und Blutgruppen; Clay Regazzoni trägt seine CARRERA.

Jack Heuer gave every Ferrari driver a golden Carrera engraved with their name and blood type on the back case; Clay Regazzoni wearing his Carrera.

TACHY
CARRERA
HEUER
AUTOMATIC
CHRONOGRAPH
SWISS

CARRERA

ref. 1153 BN „Yachting“

Carrera-Automatikgehäuse mit verschraubtem Boden, auf der Oberseite und den Seiten gebürstet. Spektakuläres blaues Zifferblatt mit schwarzen 30-Minuten- (orangefarbene 5-Minuten-Skalen) sowie 12-Stunden-Hilfszifferblättern, umrahmt von einem weißen Tachymeter-Ring mit schwarzen Ziffern. Unter 12 Uhr steht in Weiß „CARRERA“ über dem Heuer-Logo, „AUTOMATIC CHRONOGRAPH“ steht über dem Datumsfenster bei 6 Uhr.
Die neu gestalteten Stundenindizes haben in der Mitte Leuchtmasse-Einlagen, die von zwei weißen Stahlstäben umgeben sind. Das Stunden-Hilfszifferblatt zeigt die Zahlen von 1 bis 12, was auf eine spätere Version hinweist. Abgeschrägte Hilfszifferblätter zeigen konzentrische kreisförmige Rillen. Die weiß bedruckten Stahlzeiger mit Leuchtmasse-Einlagen sind in der Mitte mit einer feinen schwarzen Linie bedruckt. Der orangefarbene Chronographen-Zeiger erreicht die 1/5-Sekunden-Skala mit markanten orangefarbenen Stunden-Markierungen am äußeren Rand des Zifferblatts. Die Krone befindet sich auf der linken Seite und ist mit „HEUER“ signiert. Die Chronographen-Drücker sind auf der rechten Seite des Gehäuses integriert, aber von der Rückseite aus vollständig sichtbar. Der Gehäuseboden hat einen Rundschliff.
Die Uhr wird von einem Automatikwerk Kaliber 12 angetrieben und ist auf der Brücke signiert.

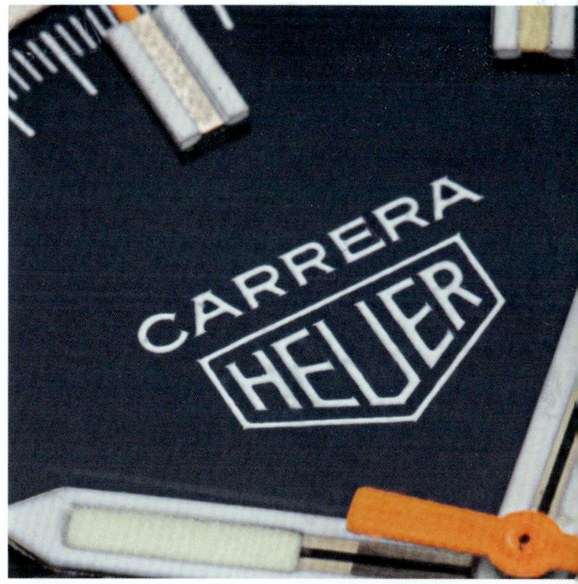

Automatic Carrera screw-back case construction with brushed finish on top and sides. Striking blue dial with black subdials for 30 minutes (orange 5-minute bars) and 12 hours, encircled by a white tachymeter ring with black numerals. Under 12 o'clock "CARRERA" is printed in white over the Heuer logo, and "AUTOMATIC CHRONOGRAPH" is above the date window at 6 o'clock. The new-design hour markers have luminous inserts at the center enclosed by two white steel bars. The hour register shows numerals from 1 to 12, indicating a later version. Angled subdials have concentric circular ridges. The white-printed steel steel hands have expanded luminous inlays and a fine black line printed down the center. The orange chronograph hand reaches the 1/5-second track with striking orange hour marks on the outer edge of the dial. The crown is on the left and signed "HEUER." Chronograph pushers are integrated on the right of the case but fully visible from the back. The case back has a circular grinding finish. The watch is powered by a Calibre 12 Automatic movement and is signed on the bridge.

CARRERA

ref. 110.253 G „small Carrera slate dial“

Carrera-Automatikgehäuse mit verschraubtem Boden und gebürsteter Oberfläche auf der Oberseite und den Seiten. Ästhetisches schieferfarbenes „Sunburst“-Zifferblatt mit anthrazitfarbenen 30-Minuten- sowie 12-Stunden-Hilfszifferblättern, eingerahmt von einem Tachymeter-Ring mit weißen Ziffern. Der Modellname Carrera wurde bei dieser seltenen Version von 1978 gegenüber dem Heuer-Logo stark verkleinert. „AUTOMATIC CHRONOGRAPH“ steht oberhalb des Datumsfensters über „SWISS“. Die eleganten Stundenindizes aus Stahl sind schwarz und haben Leuchtpunkte am äußeren Ende, die Indizes bei 3, 6 und 9 Uhr wurden weggelassen. Das Minuten-Hilfszifferblatt zeigt zum ersten mal bei der Carrera weiße Ziffern bei 5, 10, 15, 20, 25 und 30.
Die gebürsteten Stahlzeiger mit stark vergrößerten Leuchtmasse-Einlagen sind schwarz umrahmt und haben schwarze Spitzen. Der schwarze Chronographen-Zeiger erreicht die 1/5-Sekunden-Skala am äußeren Rand des Zifferblatts. Die Krone befindet sich auf der linken Seite und ist mit „HEUER“ signiert. Die Chronographen-Drücker sind auf der rechten Seite des Gehäuses integriert, aber von der Rückseite aus vollständig sichtbar. Der Stahlboden hat einen Längsschliff. Die Referenz und die Seriennummer sind auf dem äußeren Gehäuseboden eingraviert, statt zwischen den Bandanstößen.
Die Uhr wird von einem Automatikwerk Kaliber 12 angetrieben und ist auf der Brücke signiert.

Automatic Carrera screw-back case with brushed finish on top and sides. The exquisite slate starburst dial with charcoal subdials recording 30 minutes and 12 hours, encircled by a matching tachymeter ring with white numerals. The Carrera model name was substantially reduced versus the Heuer logo on this rare version from 1978. "AUTOMATIC CHRONOGRAPH" is printed above "SWISS" top of the date window. Fine steel hour markers have black printing and luminous dots at the outer end; markers have been omitted at 3, 6, and 9 o'clock. The minute register has 5, 10, 15, 20, 25, 30 indicated in white for the first time on Carreras. Brushed steel hands with enlarged luminous inserts have black printing towards the outer sides and black tips. The black chronograph hand reaches the 1/5-second track on the outer edge of the dial. The crown is on the left side and signed "HEUER." Chronograph pushers are integrated on the right of the case but fully visible from the back. The steel case back has longitudinal grinding. The reference and serial number are engraved on the outer back case instead of between the lugs.
The watch is powered by a Calibre 12 Automatic movement and is signed on the bridge.

TACHY
CARRERA
HEUER
AUTOMATIC
CHRONOGRAPH
SWISS
18

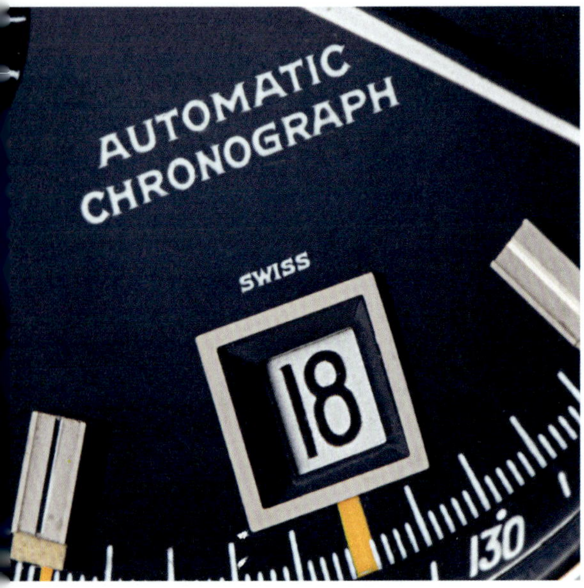

CARRERA

ref. 110.253 B „Triband-Dial"

Carrera-Automatikgehäuse mit verschraubtem Boden und gebürsteter Oberfläche an der Oberseite und den Seiten. Das revolutionäre Zifferblatt ist in Blau-Metallic gehalten. Zwei Längsstreifen unterteilen das Zifferblatt in drei Teile (Triband), die seitlich gebürstete Oberfläche sorgt für fantastische Reflexe und Lichtspiele. Ovale schwarze 30-Minuten- und 12-Stunden-Hilfszifferblätter werden von einem blauen Tachymeter-Ring mit weißen Ziffern umrahmt. Unter 12 Uhr steht in weiß „CARRERA" über dem Heuer-Logo, „AUTOMATIC CHRONOGRAPH" steht über dem Datumsfenster bei 6 Uhr. Die Stundenindizes aus Stahl haben Leuchtbalken am äußeren Ende, die Indizes bei 3, 6 und 9 Uhr wurden weggelassen.
Stabzeiger aus Stahl mit vergrößerten Leuchtmasse-Einlagen. Ein weißer Chronographen-Zeiger erreicht die 1/5-Sekunden-Skala am äußeren Rand des Zifferblatts mit markanten gelb-orangenen Stundenmarkierungen. Breitere Markierungen befinden sich auf 3, 6 ,9 und 12 Uhr. Die Krone befindet sich auf der linken Seite und ist mit „HEUER" signiert. Die Chronographen-Drücker sind auf der rechten Seite des Gehäuses integriert, aber von der Rückseite aus vollständig sichtbar. Der Stahlboden hat einen Längsschliff. Zum ersten Mal sind die Referenz- und die Seriennummer nicht zwischen den Bandanstößen, sondern auf dem äußeren Gehäuseboden eingraviert.
Die Uhr wird von einem Automatikwerk Kaliber 12 angetrieben und ist auf der Brücke mit „HEUER CAL 12" signiert.

Automatic Carrera screw-back case with brushed finish on top and sides. The revolutionary dial has a blue metallic finish. Two longitudinal bands divide the dial into three parts (triband), the lateral brushed finish creates fascinating reflections and light interplay. Oval black subdials recording 30 minutes and 12 hours are encircled by a tachymeter ring with white numerals. Under 12 o'clock, "CARRERA" is printed in white over the Heuer logo; "AUTOMATIC CHRONOGRAPH" is above the date window at 6 o'clock. Steel hour markers have luminous bars at the outer end; markers have been omitted at 3, 6, and 9 o'clock.
The steel baton hands have enlarged luminous inserts. A white chronograph hand reaches the 1/5-second track on the outer edge of the dial; there are striking yellow-orange hour marks on the outer edge of the dial, with wider marks at 12, 3, 6, and 9 o'clock.
The crown is on the left and signed "HEUER." The chronograph pushers are integrated on the right of the case but fully visible from the back. The steel case back has longitudinal grinding. For the first time, the reference and serial number are engraved on the outer back case instead of between the lugs.
The watch is powered by a Calibre 12 Automatic movement and is signed "HEUER CAL 12" on the bridge.

TACHY
CARRERA
HEUER
AUTOMATIC
CHRONOGRAPH
SWISS

VON KÖCK

TAGHeuer
SWISS AVANT-GARDE SINCE 1860

MENÜKARTE

KÖCK-TAG-HEUER-ABENDESSEN
"DIE VERGANGENHEIT TRIFFT DIE GEGENWART"

WIEN, 18.05.2010

150 YEARS PIONEERING SWISS WATCHMAKING
FOR 150 YEARS

Jack überreicht Niki eine goldene Carrera als Ersatz für die beim Einbruch gestohlene.
Jack presents Niki with a golden Carrera to replace the one stolen during the burglary.

TACHY
CARRERA
HEUER
AUTOMATIC
CHRONOGRAPH
SWISS
18

CARRERA

ref. 1158 CHN 18K gold

18-Karat-Gehäuse mit verschraubtem Boden und Strichschliff auf der Gehäuseoberseite und an den Flanken. Die Goldpunze (18 Kt.) sitzt zwischen den Chronographen-Drückern auf der rechten Seite. Das Zifferblatt ist goldfarben, mit schwarzen Hilfszifferblättern, die 30 Minuten bzw. 12 Stunden anzeigen, und das ist von einem goldenen Tachymeter-Ring mit schwarzen Zahlen umgeben. Bei 12 Uhr ist ein rechteckiger Goldindex mit schwarzem Steineinschluss. Der „CARRERA"-Modellname steht über dem Heuer-Logo und „AUTOMATIC CHRONOGRAPH" über dem weißen Datumsfenster mit weißer Zahl.
Feine goldene Parallel-Indizes haben schwarze Einlagen und Leuchtpunkte am äußeren Ende, keine Indizes bei 3, 6 und 9 Uhr. Jede einzelne Zahl des 12-Stunden-Hilfszifferblatts ist in gold gedruckt. Die Hilfszifferblätter sind abgeschrägt und haben konzentrische Rillen. Gebürstete Goldzeiger haben mit Leuchtmasse gefüllte Flächen, sowie schwarze Umrahmungen und Spitzen. Der schwarze Chronographen-Zeiger reicht präzise zu der 1/5-Sekunden-Skala. Die Krone ist mit „HEUER" signiert. Die Drücker sind im Gehäuse integriert. Der innere Gehäuseboden ist mit einem Rundschliff und einer Goldpunze versehen.
Die Uhr wird angetrieben von einem Kaliber 12 Automatik-Uhrwerk und ist auf der Brücke mit „HEUER CAL.12" signiert.
Diese Uhr wurde von Gerd-Rüdiger Lang, einem ehemaligen Mitarbeiter und Gründer seiner eigenen Marke Chronoswiss, direkt bei Heuer neu bestellt. Er war langjähriger Mitarbeiter bei Heuer und ein Pionier der unabhängigen Uhrmacherei.

18-karat gold screw-back case with brushed finish on top and sides. A hallmark is between the chronograph pushers at the right. The dial is gold with black subdials recording 30 minutes and 12 hours and is surrounded by a gold tachymeter ring with black numerals. At 12 o'clock there is a beautiful rectangular gold index with a black stone insert. "CARRERA" is shown over the Heuer logo and "AUTOMATIC CHRONOGRAPH" is printed above the white date window with white font.
The fine twin gold hour markers have black inserts and luminous dots at the outer end; markers have been omitted at 3, 6, and 9 o'clock. The hour register has each individual number painted in gold. The subdials are angled, with flat faces with concentric circular ridges. The brushed gold hands with luminous material have black printing towards the outer sides and black tips. The black chronograph hand reaches the 1/5-second track on the outer edge of the dial. The crown is signed "HEUER." The pushers are integrated. The inner case back has circular grinding and gold hallmarks.
The watch is powered by a Calibre 12 Automatic movement and is signed "HEUER CAL.12" on the bridge.
This one was ordered new directly from Heuer by Gerd-Rüdiger Lang, a former employee and founder of his own brand Chronoswiss. A pioneer of independent watchmaking.

BASE 30 PULSATIONS
HEUER
CARRERA
AUTOMATIC
CHRONOGRAPH
SWISS
18

CARRERA

ref. 1553 N „Pulsometer“

Erste Ausführung des verschraubten Carrera Automatik-Gehäuses; gebürstetes Finish auf der Oberseite und den Seiten. Das Zifferblatt ist leuchtend blau mit einer kleinen permanenten Sekunde bei 10 Uhr, direkt darunter befindet sich das „CARRERA“-Logo, das bei dieser Variante neben dem 9-Uhr-Index auf einem silbernen Balken positioniert ist, der zum silbernen 30-Minuten-Hilfszifferblatt bei 3 Uhr führt. Das Zifferblatt wird von einer sehr seltenen weißen Pulsometer-Lünette mit blauer Skala umrahmt. Das Heuer-Logo befindet sich unterhalb von 12 Uhr, über dem Datumsfenster steht „AUTOMATIC CHRONOGRAPH“.
Die weiß lackierten Stundenindizes aus Stahl sind mit großen Leuchtmasse-Einlagen versehen. Die Indizes bei 3 und 6 Uhr wurden weggelassen, dafür gibt es winzige Leuchtpunkte. Das Hilfszifferblatt mit den üblichen 10-, 20- und 30-Minuten-Markierungen ist nicht vertieft.
Die weiß lackierten Stahlzeiger sind mit Leuchtmasse und feinen schwarzen Einsätzen in der Mitte versehen. Der orangefarbene Chronographen-Zeiger erreicht die 1/5-Sekunden-Skala am äußeren Rand des Zifferblatts.
Die Krone befindet sich auf der linken Seite und ist mit „HEUER“ signiert. Die Chronographen-Drücker sind auf der rechten Seite des Gehäuses integriert, aber von der Rückseite aus vollständig sichtbar. Der Gehäuseboden dieser ungetragenen Uhr ist mit einem roten Heuer-Aufkleber versehen.
Die Uhr wird von einem Automatikwerk Kaliber 15 angetrieben und ist bei dieser unpolierten Carrera auf der Brücke mit „ CAL 15“ signiert.

First-version Automatic Carrera screw-back case with brushed finish on top and sides. The magnificent dial is bright blue with a small permanent second hand at 10 o'clock, just below is the "CARRERA" logo now repositioned aside the 9 o'clock marker on a silver bar leading to the silver 30-minute subdial at 3 o'clock.
The dial is encircled by a very rare white pulsometer bezel with blue printed scale. The Heuer shield logo remains under 12 o'clock and "AUTOMATIC CHRONOGRAPH" is printed above the date window. White-painted steel hour markers have large luminous inserts. Markers have been omitted at 3 and 6 o'clock but there are tiny lume dots. The sub dial is not recessed and has the usual 10-, 20-, and 30-minute markers. The white-painted steel hands have luminous and fine black inserts towards the centre. The orange chronograph hand reaches the 1/5-second track on the outer edge of the dial. The crown is on the left and signed "HEUER." Chronograph pushers are integrated on the right of the case but fully visible from the back. The case back of this unworn new old-stock watch has a red Heuer sticker.
The watch is powered by a Calibre 15 Automatic movement and is signed "CAL 15" on the bridge on this unpolished original survivor.

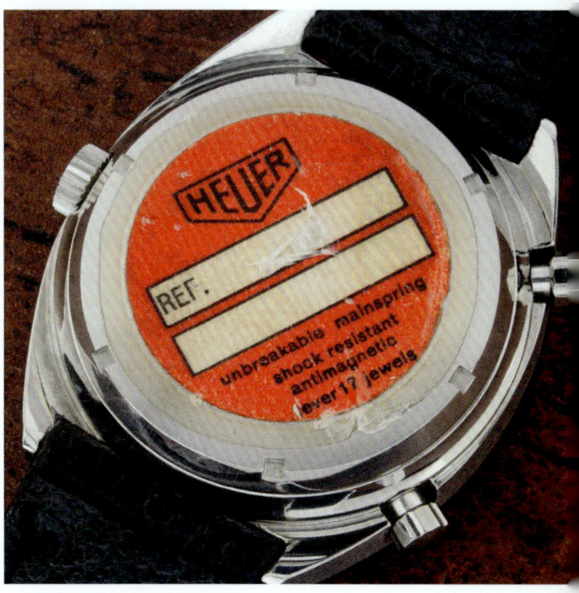

HEUER
REF.
unbreakable mainspring
shock resistant
antimagnetic
lever 17 jewels

CARRERA

ref. 110.573 F „Triband-Dial“

Das 1974 neu eingeführte, avantgardistische Carrera-Automatikgehäuse (wegen seiner Dicke auch „Barrel-Case“ (dt. „Fass“) genannt) ist oben gebürstet und an den Seiten poliert. Das innovative Zifferblatt ist in Braunmetallic ausgeführt. Zwei längs verlaufende Streifen teilen das Zifferblatt in drei Teile. Die Seiten sind gebürstet, was zu fantastischen Reflexen und Lichtspielen führt. Die anthrazitfarbenen 30-Minuten- und 12-Stunden-Hilfszifferblätter sind oval geformt (in Anlehnung an die ovale Lünette an der Gehäuseoberseite).
Das Zifferblatt ist von einem stark abgeschrägten, braunen Tachymeter-Ring aus Metall umgeben. Unterhalb von 12 Uhr steht „CARRERA“ über dem Heuer-Logo, und über dem Datumsfenster bei 6 Uhr ist „AUTOMATIC CHRONOGRAPH“ aufgedruckt. Die Stundenindizes aus Metall haben Leuchtbalken am äußeren Ende, die Indizes bei 3, 6 und 9 Uhr wurden weggelassen. Die gelb lackierten Stahlzeiger haben größere Leuchtmittel-Einlagen. Der gelbe Chronographen-Zeiger erreicht die 1/5-Sekunden-Skala am äußeren Rand des Zifferblatts. Die Krone befindet sich auf der linken Seite und ist mit „HEUER“ signiert. Die Chronographen-Drücker sind in die Krone integriert.
Am Gehäuseboden dieser ungetragenen Uhr klebt ein original Heuer-Aufkleber.
Die Uhr wird von einem Kaliber 12 Automatikwerk angetrieben und ist auf der Brücke signiert.
Diese ungetragene Uhr hat ein originales Stahlband.

Newly introduced for 1974, firmly modernist automatic Carrera screw-back case (known as a "barrel case" due to its thickness) with brushed finish on top and polished sides. The innovative dial has a brown metallic finish. Two longitudinal bands divide the dial into three parts; the lateral brushed finish creates fascinating reflections and light interplay. Anthracite subdials have a new oval shape (reflecting the oval bezel of the top case), recording up to 30 minutes and 12 hours.
The dial is surrounded by a deeply beveled brown metallic tachymeter ring. Below 12 o'clock "CARRERA" stands over Heuer logo and automatic chronograph" is printed above the date window at 6 o'clock. Metal hour markers have luminous bars at the outer end; markers have been omitted at 3, 6, and 9 o'clock. Yellow painted steel hands have larger faces for luminous materials. Yellow chronograph hand reaches the 1/5-second track on the outer edge of the dial. The crown is on the left and signed "HEUER." The chronograph pushers are integrated on the right of the case but fully visible from the back. The case back has an original Heuer sticker on this unworn new old-stock example.
The watch is powered by a Calibre 12 Automatic movement and is signed on the bridge. An original steel band is attached.

TACHY
CARRERA
HEUER
AUTOMATIC
CHRONOGRAPH
SWISS
18

CARRERA

ref. 110.573 B „Angola“

Modernistisches Carrera Automatik-Gehäuse mit verschraubtem Boden (aufgrund seiner Dicke auch „Barrel-Case“ genannt) mit gebürstetem Finish auf der Oberseite und polierten Seiten. Das avantgardistische Zifferblatt ist blau-metallic. Zwei Längsstreifen unterteilen es in drei Teile. Die gebürstete Oberfläche führt zu fantastischen Reflexen und Lichtspielen. Die beiden anthrazitfarbenen 30-Minuten- und 12-Stunden-Hilfszifferblätter sind oval geformt (in Anlehnung an die ovale Lünette).
Das dreidimensionale Zifferblatt ist von einem stark abgeschrägten, blau-metallenen Tachymeter-Ring umgeben. Unterhalb von 12 Uhr steht „CARRERA“ über dem Heuer-Logo, und über dem Datumsfenster bei 6 Uhr ist „AUTOMATIC CHRONOGRAPH“ aufgedruckt. Die Stundenindizes aus Metall haben Leuchtbalken am äußeren Ende, die Indizes bei 3, 6 und 9 Uhr wurden weggelassen. Die gelb lackierten Stahlzeiger haben größere Leuchtmasse-Einlagen. Der gelbe Chronographen-Zeiger erreicht die 1/5-Sekunden-Skala am äußeren Rand des Zifferblatts. Die Krone befindet sich auf der linken Seite und ist mit „HEUER“ signiert. Die Chronographen-Drücker sind auf der rechten Seite des Gehäuses integriert, aber von der Rückseite aus vollständig sichtbar. Der Gehäuseboden dieser seltenen Militäruhr wurde für die angolanische Armee graviert.
Die Uhr wird von einem Automatikwerk Kaliber 12 angetrieben und ist auf der Brücke signiert.

Firmly modernist automatic Carrera screw-back case (known as "barrel case" due to its thickness) with brushed finish on top and polished sides. The leading-edge dial has a blue metallic finish. Two longitudinal bands divide the dial into three parts; the lateral brushed finish creates fascinating reflections and light interplay. The charcoal subdials are in the new oval shape (reflecting the oval bezel of the top case), recording up to 30 minutes and 12 hours.
The dial is surrounded by a deeply beveled blue metallic tachymeter ring. Below 12 o'clock the "CARRERA" name is over the Heuer logo and "AUTOMATIC CHRONOGRAPH" is printed above the date window at 6 o'clock. The metal hour markers have luminous bars at the outer end; markers have been omitted at 3, 6, and 9 o'clock. The yellow-painted steel hands have large luminous inlays. The yellow chronograph hand reaches the 1/5-second track on the outer edge of the dial. The crown is on the left and signed "HEUER." The chronograph pushers are integrated on the right of the case but fully visible from the back. The case back on this rare military watch was engraved for the Angolan Armed Forces.
The watch is powered by a Calibre 12 Automatic movement and is signed on the bridge.

TACHY
CARRERA
HEUER
AUTOMATIC
CHRONOGRAPH
SWISS

CARRERA

ref. 110.515 CHN

Das 1974 neu eingeführte, vergoldete Carrera-Automatikgehäuse mit verschraubtem Boden (wegen seiner Dicke auch Barrel-Case genannt) ist auf der Oberseite gebürstet und an den Seiten poliert. Das außergewöhnliche Zifferblatt ist champagnerfarben. Zwei längs verlaufende Streifen teilen das Zifferblatt in drei Teile. Die gebürstete Oberfläche bietet beeindruckende Reflexionen. Die schwarzen 30-Minuten- und 12-Stunden-Hilfszifferblätter sind oval (in Anlehnung an die ovale Lünette des oberen Gehäuses).
Das Zifferblatt ist von einem stark abgeschrägten goldfarbenen Tachymeter-Ring umgeben. Unterhalb von 12 Uhr steht „CARRERA" über dem Heuer-Logo und „AUTOMATIC CHRONOGRAPH" über dem Datumsfenster bei 6 Uhr. Die goldenen Stundenindizes haben Leuchtbalken am äußeren Ende, die Indizes bei 3, 6 und 9 Uhr wurden weggelassen.
Die schwarz lackierten und goldgebürsteten Zeiger haben größere Leuchtmasse-Einlagen. Der schwarze Chronographen-Zeiger erreicht die 1/5-Sekunden-Skala am äußeren Rand des Zifferblatts. Die goldene Krone befindet sich auf der linken Seite und ist mit „HEUER" signiert. Die vergoldeten Chronographen-Drücker sind auf der rechten Seite des Gehäuses integriert, aber von der Rückseite aus vollständig sichtbar. Der Stahlboden hat einen Längsschliff.
Die Uhr wird von einem Automatikwerk Kaliber 12 angetrieben und ist auf der Brücke signiert.

Newly introduced for 1974, gold-plated automatic Carrera screw-back case (known as "barrel case" due to its thickness) with brushed finish on top and polished sides. The modernist dial ihas a champagne finish. Two longitudinal bands divide the dial into three parts; the lateral brushed finish creates fascinating reflections and light interplay. Black subdials have a new oval shape (reflecting the oval bezel of the top case), recording up to 30 minutes and 12 hours. The dial is surrounded by a deeply beveled gold tachymeter ring. Below 12 o'clock the "CARRERA" name is over the Heuer logo and "AUTOMATIC CHRONOGRAPH" is printed above the date window at 6 o'clock. Golden hour markers have luminous bars at the outer end; markers have been omitted at 3, 6, and 9 o'clock. The black-painted and gold-brushed hands have larger faces with luminous inlays. The gold-plated chronograph hand reaches the 1/5-second track on the outer edge of the dial. The crown is on the left and Signed "HEUER." Plated chronograph pushers are integrated on the right of the case but fully visible from the back. The steel case back has longitudinal grinding.
The watch is powered by a Calibre 12 Automatic movement and is signed on the bridge.

CARRERA

ref. 110.571 NC

Abgebildet im Heuer-Katalog von 1977. Schwarzes Carrera-Automatikgehäuse (wegen seiner Dicke auch „Barrel-Case" genannt). Das Zifferblatt ist schwarz, mit kontrastierendem weißem Aufdruck, sowie 30-Minuten- und 12-Stunden-Hilfszifferblättern.
Das Zifferblatt ist von einem stark abgeschrägten schwarzen Tachymeter-Ring umgeben. Unterhalb von 12 Uhr steht „CARRERA" über dem Heuer-Logo und „AUTOMATIC CHRONOGRAPH" über dem Datumsfenster bei 6 Uhr. Die Stundenindizes bestehen aus schmalen, vollständig mit Leuchtmasse belegten Balken, die Indizes bei 3, 6 und 9 Uhr wurden weggelassen. Die weiß lackierten Zeiger haben größere Leuchtmasse-Einlagen. Ein weißer Chronographen-Zeiger erreicht die 1/5-Sekunden-Skala am äußeren Rand des Zifferblatts. Die Stahlkrone befindet sich auf der linken Seite und ist mit „HEUER" signiert. Die Chronographen-Drücker aus Stahl sind auf der rechten Seite des Gehäuses integriert, aber von der Rückseite aus vollständig sichtbar. Das Gehäuseoberteil ist schwarz beschichtet, der Gehäuseboden aus Stahl.
Die Uhr wird von einem Automatikwerk Kaliber 12 angetrieben und ist auf der Brücke mit „HEUER CAL.12" signiert.

Shown in the 1977 Heuer catalogue. Black finished automatic Carrera screw-back case (known as "barrel case" due to its thickness). The dial is black with high-contrast white print; subdials record up to 30 minutes and 12 hours.
The dial is encircled by a deeply beveled black tachymeter ring. Below 12 o'clock, "CARRERA" is printed over the Heuer logo and "AUTOMATIC CHRONOGRAPH" above the date window at 6 o'clock. The slim hour markers are fully luminous bars; markers have been omitted at 3, 6, and 9 o'clock. The white painted hands have large luminous inlays. The white chronograph hand reaches the 1/5-second track on the outer edge of the dial. The steel crown is on the left and Signed "HEUER." Steel chronograph pushers are integrated on the right of the case but fully visible from the back. The top case is fully coated black, while the case back is steel.
The watch is powered by a Calibre 12 Automatic movement and is signed "HEUER CAL. 12" on the bridge.

TACHY
CARRERA
HEUER
AUTOMATIC
CHRONOGRAPH
SWISS
18

τηλ. 22257 - ρόδος
CORTINA
HEUER
SWISS

CORTINA

Der Name „Cortina“ weckt zwei Assoziationen. Erstens verbindet man damit den weltbekannten Wintersporttreffpunkt der Reichen, Eleganten und Berühmten in den Dolomiten im Norden Italiens und Austragungsort der Olympischen Winterspiele von 1956. Das Rennen in Cortina ist ein fester Bestandteil des alljährlichen Skiweltcup. Zweitens denkt man an den Motorsport und den Lotus Cortina. Jim Clark und Vic Elford steuerten die schnellen Cortinas im Tourenwagen- und im Rallye-Sport Mitte der 1960er Jahre.

Jack Heuer war sicher von dem internationalen Publikum begeistert, das sich jedes Jahr in Cortina trifft. Eine wichtige Assoziation für die Firma Heuer war der Wintersport, hatte man sich doch als Zeitmesser bei verschiedenen Weltcupveranstaltungen schon eindrucksvoll einen Namen gemacht.

Somit war die Wahl eines Wintersportortes für die neue Modellreihe naheliegend. Cortina stand für Eleganz, beste Restaurants, tolle Geschäfte und die weltberühmte Bobbahn. Die Begeisterung für den Skisport war in Italien und den angrenzenden Ländern enorm, und im Portfolio gab es bislang kein Modell mit klarer Assoziation zum Wintersport, obwohl man in diesem Sportsegment so aktiv war. Jack Heuer beschloss, diesen White Spot auszumerzen, und nannte seine neue Chronographen-Linie Mitte der 70er Jahre „CORTINA“.

Ein Novum im Portfolio ist das weiß marmorierte Zifferblatt mit römischen Ziffern und elegant geformtem, schwarz lackiertem Zeigerspiel. Der Modellname „CORTINA“ ist über dem Heuer-Logo im oberen Bereich des Blattes platziert. Angetrieben durch das bewährte Kaliber 12 Automatik-Werk, stellt das Gehäuse ein Novum dar. Das integrierte Stahlband schuf eine elegante Verbindung zum mit „HEUER“ signierten Verschluss. Auch ein schwarzes Zifferblatt wurde in den Katalogen zwischen 1977 und 1979 angeboten.

The name "Cortina" evokes two significant associations. First, it is associated with the famous winter sports town for the rich and famous in the Dolomite Mountains of northern Italy, where the 1956 Winter Olympics took place and the ski competition is held every year during the Ski World Cup. The second association is the connection to motorsports and the Lotus Cortina. Jim Clark and Vic Elford drove the fast Cortinas in touring car and rally sport races during the mid-1960s.

Jack Heuer was certainly excited by the international scene that met every winter in Cortina. An important association for the Heuer company was the field of winter sports, for which it had provided timekeeping during the various World Cup events, building an excellent reputation.

It was thus logical to choose a winter sports center for the name of a new model line. Cortina stood for chic, the best restaurants, fantastic stores, and the famous bobsled track. There was enormous enthusiasm for skiing in Italy and its neighboring countries. Heuer, however, still lacked a model in its portfolio which reflected a clear connection with winter sports, the company's extensive activities in this field. Jack Heuer thus decided to fill in the blanks and called his new line of chronographs, released in the late 1970s, "CORTINA."

A new addition was a white marbled dial with Roman numerals and elegant, gleaming black hands. The model name "CORTINA" was placed above the Heuer logo on the dial. The watch is powered by the proven Calibre 12 automatic movement. The integrated steel watchband provided an elegant, smooth transition up to the Heuer-signed clasp. A second version with a black dial was advertised in the catalogs between 1977 and 1979.

CHRONOGRAPH
HEUER
CORTINA
HEUER
AUTOMATIC
CHRONOGRAPH
SWISS
18
CHRONOGRAPH
HEUER

CORTINA

ref. 110.233 R

Sechskantiges Gehäuse mit verschraubtem Boden, Strichschliff, einem polierten Ring um das Glas und polierten Seiten. Das Stahlband ist speziell für die Cortina-Modelllinie gefertigt und ist mit dem Gehäuse beweglich verbunden. Das glänzend weiße Zifferblatt hat römische Ziffern mit 30-Minuten- und 12-Stunden-Hilfszifferblättern. Der Modellname „CORTINA" steht über dem Heuer-Logo unter 12 Uhr. „AUTOMATIC CHRONOGRAPH" steht über dem schwarz gerahmten Datumsfenster bei 6 Uhr.
Die glänzend schwarzen römischen Zahlen der Stundenindizes sind am äußeren Ende des weißen Zifferblattes innerhalb der Sekunden-Skala platziert. Die Zeiger sind glänzend schwarz und ohne Leuchtmasse. Der glänzend schwarze Chronographen-Sekundenzeiger berührt die schwarze 1/5-Sekunden-Skala am äußeren Rand des Zifferblattes. Die Krone ist links mit „HEUER" signiert. Die Drücker sind rechts. Der Rückdeckel mit Längsschliff und polierten Seiten hat die Referenz- und Seriennummer eingraviert.
Das Uhrwerk ist ein Kaliber 12 und mit „CAL. 12" auf der Brücke signiert.

Hexagonal automatic screw-back case with brushed finish, polished circle around the crystal, and polished sides. The steel bracelet was custom built for the Cortina model range and is pin-jointed to the case. The luminous white dial has Roman numerals and 30-minute and 12-hour recording subdials. The model name "CORTINA" is positioned over the Heuer logo below 12 o'clock. "AUTOMATIC CHRONOGRAPH" is above the black-framed date window at 6 o'clock.
The glossy black Roman numerals serve as hour markers and are placed on the outer edge of the white dial inside the second track. Numerals have been omitted at 3, 6, and 9 o'clock. The slim hands are glossy black without luminous inserts. The glossy black chronograph second hand reaches the black 1/5-second track on the outer edge of the dial. The crown is on the left and signed "HEUER" and the pushers are the right. The case back is engraved with the reference and serial number and has longitudinal grinding and polished sides.
The movement is a Caliber 12, and is signed "CAL. 12" on the bridge.

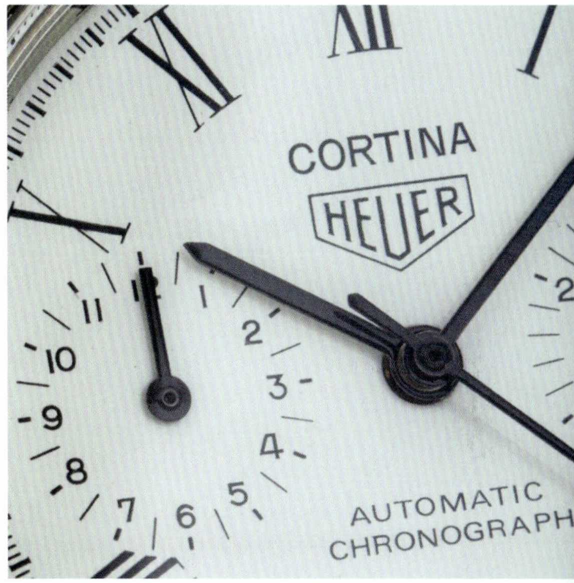

HEUER

Jack Heuer inspiziert die Funktion des 1/1000 Sekunden anzeigenden Microtimers, der bei Weltcup-Skirennen eine genaue Zeitmessung ermöglichte.

Jack Heuer examines the function of the microtimers displaying 1/1000-second increments, which allowed accurate timing at Ski World Cup competitions.

ROLEX
HEUER - CAMY
Διέθεσεν
Διεύθυνση κατοικίας
CORTINA
HEUER
AUTOMATIC
CHRONOGRAPH
SWISS

CORTINA

ref. 110.233 NC

Sechskantiges Gehäuse mit verschraubtem Boden, Strichschliff, einem polierten Ring um das Glas und polierten Seiten. Das Stahlband ist speziell für die Cortina-Modelllinie gefertigt und mit dem Gehäuse beweglich verbunden. Das schwarze Zifferblatt hat ein 30-Minuten- und 12-Stunden-Hilfszifferblatt. Der Modellname „CORTINA" steht über dem Heuer-Logo unter 12 Uhr. „AUTOMATIC CHRONOGRAPH" steht über dem weiß gerahmten Datumsfenster bei 6 Uhr.

Die mit Leuchtmasse gefüllten Stundenindizes sind am äußeren Ende des schwarzen Zifferblattes innerhalb der Sekunden-Skala platziert. Ein Doppelindex steht bei 12 Uhr, kleinere quadratische Indizes bei 3, 6 und 9 Uhr. Die weißen Zeiger haben Leuchtmaterial-Einlagen. Der weiße Chronographen-Sekundenzeiger berührt die weiße 1/5-Sekunden-Skala am äußeren Rand des Zifferblattes. Die Krone ist links mit „HEUER" signiert. Die Drücker sitzen rechts. Der Rückdeckel mit Längsschliff und polierten Seiten hat die Referenz- und Seriennummer eingraviert. Das Uhrwerk ist ein Kaliber 12 und mit „CAL. 12" auf der Brücke signiert.

Hexagonal automatic screw-back case with brushed finish, polished circle around the crystal, and polished sides. The steel bracelet was custom-built for the Cortina model range and is pin-jointed to the case. The black dial has 30-minute and 12-hour recording subdials. The model name "CORTINA" is positioned over the Heuer logo below 12 o'clock. "AUTOMATIC CHRONOGRAPH" is above the white framed date window at 6 o'clock.

Luminous hour markers are on the outer edge of the black dial inside the second track. A twin marker is at 12 o'clock with smaller luminous dots at 3, 6, and 9 o'clock. The white hands have luminous inserts. The white chronograph second hand reaches the white 1/5-second track on the outer edge of the dial. The crown is on the left of the case and signed "HEUER" and the pushers are on the right. The case back is engraved with the reference and serial number and has longitudinal grinding and polished sides. The movement is a Caliber 12, and is signed "CAL. 12" on the bridge.

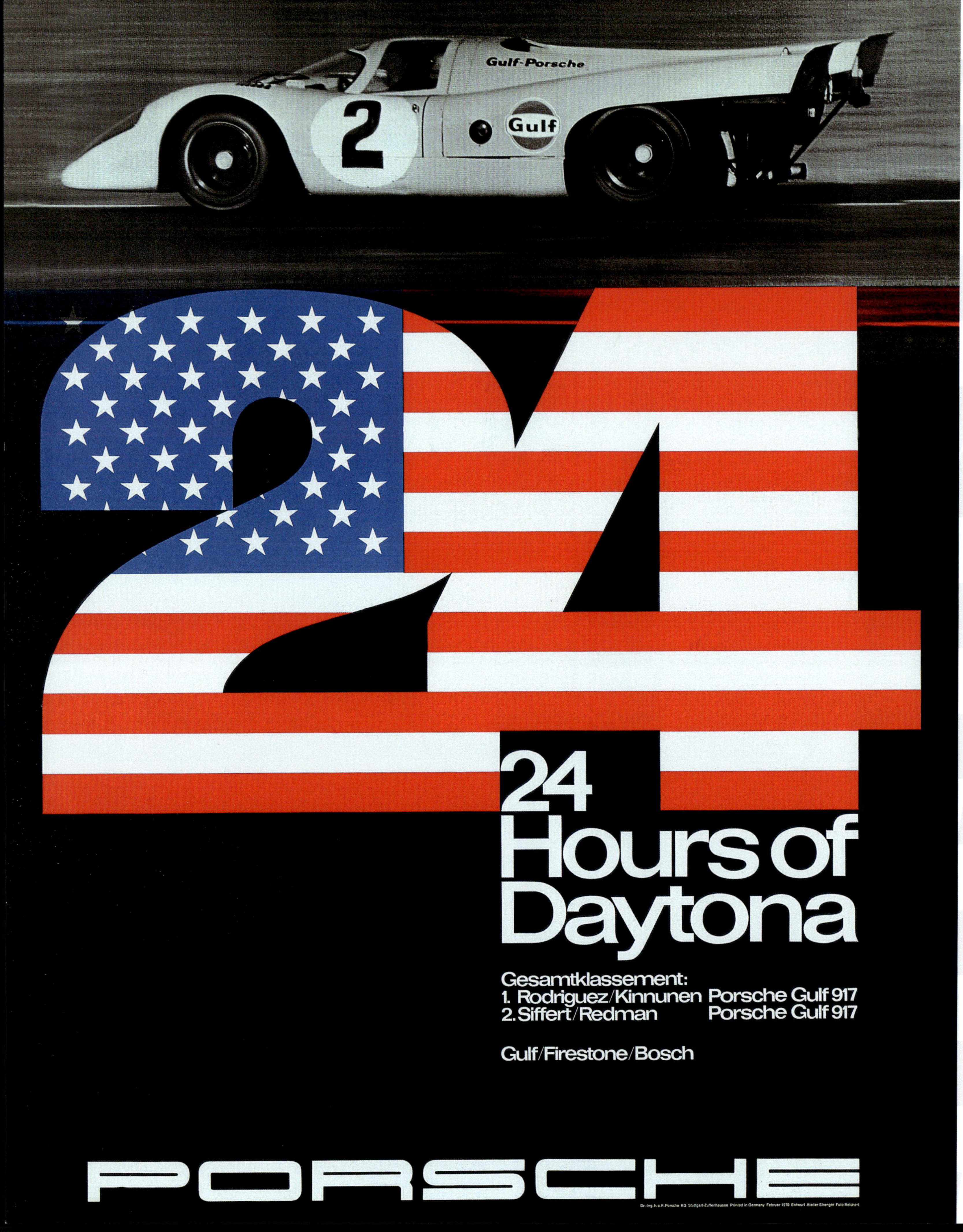

Gulf-Porsche
2
Gulf
24
24
Hours of
Daytona
Gesamtklassement:
1. Rodriguez/Kinnunen Porsche Gulf 917
2. Siffert/Redman Porsche Gulf 917
Gulf/Firestone/Bosch
PORSCHE
Dr.-Ing. h. c. F. Porsche KG Stuttgart-Zuffenhausen Printed in Germany Februar 1970 Entwurf Atelier Strenger Foto Reichert

DAYTONA

Jack Heuer war sicher von der bestechenden Eleganz des Ferrari 365GTB/4 Daytona begeistert. Wie bei Ferrari üblich, zeichnete Sergio Pininfarina für das Design verantwortlich. Den Namen „Daytona" bekam das schnelle Frontmotor-Grand-Turismo-Coupé von den großen Erfolgen bei dem berühmten 24-Stunden-Rennen in Daytona Beach in den USA.

Deshalb war der Modellname „Daytona" mit einem positiven Image behaftet und stand für sportliche Ausdauer und technischen Wettkampf auf höchstem Niveau. Die Rennsportbegeisterung der Amerikaner kannte kaum Grenzen, und auch die amerikanische Motorindustrie benannte ihre schnellsten und innovativsten Produkte mit diesem Namen. Beispiele sind der Dodge Charger Daytona, der 1969 mit enormem Heckflügel auf sich aufmerksam machte, und nicht zu vergessen die schnellen Shelby Daytonas, die Mitte der 60er Jahre auf den Pisten siegreich waren. Anfang der 70er dominierte der Porsche 917 das 24-Stunden-Rennen von Daytona.

1976 kam die Heuer-Daytona mit einem integrierten Stahlband, sie war nicht mit einem Lederband erhältlich. Das Zifferblatt zeigte eine Farbschattierung von hellerem Blau in der Mitte zu dunklerem Blau am Rand. Der Modell-Name „DAYTONA" ist über das Heuer-Logo im oberen Bereich des Blattes gesetzt. Von dessen Außenrand steigt ein schräg gestellter Tachymeter-Ring zum Gehäuserand an und vergrößert augenfällig das Zifferblatt. Angetrieben durch das bewährte Kaliber 12-Automatik-Werk, stellt das Gehäuse ein Novum dar. Das integrierte Stahlband schuf eine elegante Verbindung zum mit „HEUER" signierten Verschluss. Ein tiefblaues Zifferblatt wurde in den Katalogen zwischen 1977 bis 1979 angeboten, eine weitere Variante hatte ein bräunliches Zifferblatt.

Jack Heuer was a fan of the elegant Ferrari 365GTB/4 Daytona. As usual at Ferrari, Sergio Pininfarina was responsible for the design. The name "Daytona" for the fast front-motor Gran Turismo coupé came from the model's standout successes in the famous 24-hour race at Daytona Beach in the USA.

This profile provided the model name "Daytona" with a an excellent image, symbolizing sporting endurance and technical competition at the highest level. The USA showed boundless enthusiasm for motorsport, and the US. Motor industry also used the name for their products. Examples are the 1969 Dodge Charger Daytona, in striking design with enormous fins, and the unforgettably fast Shelby Daytona, which won numerous races in the mid-1960s. All of them bore the "Daytona" name. At the beginning of the 70s, the mighty Porsche 917 dominated the 24 Hours of Daytona.

1976 saw the Heuer Daytona, with an integrated steel watchband. It was not available with a leather band. The dial featured color gradation from lighter blue in the center to dark blue at the edges. The model name "DAYTONA" stood above the Heuer logo on the upper part of the dial. A beveled tachymeter ring rises from the outer edge of the dial toward the case and optically increases the dial.

Powered by the proven Calibre 12 automatic movement, the case design was completely new. The integration of the steel watchband with the case was achieved by a rectangular link with a counterpart in the band itself. A deep blue dial was advertised in the catalog from 1977–1979, with brown as an alternative colorway.

Porsche-Plakat, 1970.

Porsche poster, 1970.

TACHY
DAYTONA
HEUER
AUTOMATIC
CHRONOGRAPH
SWISS
18

DAYTONA

ref. 110.203 B

Abgerundetes Gehäuse mit verschraubtem Boden und Strichschliff. Das Stahlband ist speziell für die Daytona-Modelllinie gefertigt und mit dem Gehäuse beweglich verbunden. Das metallisch blaue Zifferblatt hat 30-Minuten- und 12-Stunden-Hilfszifferblätter, umgeben von einem dunkelblauen Tachymeter-Ring. Der Modellname „DAYTONA" steht über dem Heuer-Logo unter 12 Uhr. „AUTOMATIC CHRONOGRAPH" über dem weiß gerahmten Datumsfenster bei 6 Uhr.
Die mit Leuchtmasse gefüllten Stundenindizes sind am äußeren Ende des Zifferblattes innerhalb der Sekunden-Skala platziert. Ein Doppelindex steht bei 12 Uhr, keine Indizes bei 3, 6 und 9 Uhr. Die beiden Zeiger haben große Leuchtmasse-Einlagen. Der weiße Chronographen-Sekundenzeiger berührt die weiße 1/5-Sekunden-Skala am äußeren Rand des Zifferblattes. Die Krone ist auf der linken Seite des Gehäuses integriert und mit „HEUER" signiert. Die Chronographen-Drücker sind rechts integriert. Im Rückdeckel mit Längsschliff und polierten Seiten ist die Referenz- und Seriennummer eingraviert.
Das Uhrwerk ist ein Kaliber 12 und mit „CAL. 12" auf der Brücke signiert.

Rounded screw-back case with brushed finish. The steel band was custom-built for the Daytona model series and is pin-jointed to the case. The metallic blue dial has 30-minute and 12-hour recording subdials surrounded by a dark blue tachymeter ring. The model name "DAYTONA" is positioned over the Heuer logo below 12 o'clock. "AUTOMATIC CHRONOGRAPH" is above the white-framed date window at 6 o'clock.
Luminous hour markers are along the outer edge of the dial within the second track. A twin marker is at 12 o'clock; markers have been omitted at 3, 6, and 9 o'clock. The white-painted hands have large luminous inserts. The white chronograph hand reaches the white 1/5-second track on the outer edge of the dial. The crown is integrated on the left of the case and signed "HEUER." Chronograph pushers are integrated on the right. The reference and serial numbers are engraved on the case back, which has longitudinal grinding and polished sides.
The movement is a Calibre 12, and is signed "CAL.12" on the bridge.

DAYTONA

ref. 110.213 G

Abgerundete Gehäuse mit verschraubtem Boden und Strichschliff. Das Stahlband ist speziell für die Daytona-Modelllinie gefertigt und mit dem Gehäuse beweglich verbunden. Das metallisch braune Zifferblatt hat 30-Minuten- und 12-Stunden-Hilfszifferblätter, umgeben von einem Tachymeter-Ring. Der Modellname „DAYTONA" steht über dem Heuer-Logo unter 12 Uhr. „AUTOMATIC CHRONOGRAPH" steht über dem weiß gerahmten Datumsfenster bei 6 Uhr. Die mit Leuchtmasse gefüllten Stundenindizes sind am äußeren Rand des Zifferblattes innerhalb der Sekunden-Skala platziert. Ein Doppelindex steht bei 12 Uhr, keine Indizes bei 3, 6 und 9 Uhr. Die Zeiger haben große Leuchtmasse-Einlagen. Der Chronographen-Sekundenzeiger berührt die weiße 1/5-Sekunden-Skala am äußeren Rand des Zifferblattes. Die Krone ist auf der linken Seite des Gehäuses integriert und mit „HEUER" signiert. Die Chronographen-Drücker sind rechts integriert. Der Rückdeckel hat auf der Außenseite die Referenz- und Seriennummer eingraviert, Längsschliff und polierte Seiten.
Das Uhrwerk ist ein Kaliber 12 und mit „CAL. 12" auf der Brücke signiert.

Rounded automatic Daytona screw-back case with brushed finish. The steel band was custom-built for the Daytona model series and is pin-jointed to the case. The metallic brown dial has 30-minute and 12-hour recording subdials surrounded by a dark brown tachymeter ring. The model name "DAYTONA" is positioned over the Heuer logo below 12 o'clock. "AUTOMATIC CHRONOGRAPH" is above the white-framed date window at 6 o'clock.
Luminous hour markers are along the outer edge of the dial within the second track. A twin marker is at 12 o'clock; markers have been omitted at 3, 6, and 9 o'clock. The beige-painted hands have large luminous inserts. The chronograph second hand reaches the white 1/5-second track on the outer edge of the dial. The crown is integrated on the left of the case and signed "HEUER." Chronograph pushers are integrated on the right. The reference and serial number are engraved on the case back, which has longitudinal grinding and polished sides.
The movement is a Caliber 12, and is signed "CAL.12" on the bridge.

TACHY
DAYTONA
HEUER
AUTOMATIC
CHRONOGRAPH
SWISS
18

HEUER
JARAMA
HEUER
AUTOMATIC CHRONOGRAPH
HEUER
JARAMA
HEUER
AUTOMATIC CHRONOGRAPH
HEUER
JARAMA
HEUER
AUTOMATIC CHRONOGRAPH

JARAMA

„Circuito del Jarama" ist eine Motorsport-Rennstrecke bei San Sebastian los Reyes, 32 Kilometer nördlich der spanischen Hauptstadt Madrid. Der 3,4 km lange Rundkurs war zwischen 1968 und 1981 neun Mal Austragungsort des spanischen Formel 1 Grand Prix.

Aber wie kam es zur Namensgebung der Heuer Jarama? Nach Niki Lauda's schwerem Unfall 1976 auf dem Nürburgring und dem daraus resultierenden Verlust der Weltmeisterschaft um einen Punkt an James Hunt, verlief die neue Saison 1977 für den Österreicher und die Scuderia richtig gut. Am Ende reichte Niki Lauda schon vorzeitig ein vierter Platz in Watkins Glen (USA) zu seinem zweiten Weltmeistertitel in der Königsklasse.

Ich habe Jack Heuer und Niki Lauda bei einem Treffen in Wien befragt, wie es denn zum Namen Heuer Jarama gekommen sei. Für Niki war es logischerweise sein besonderer Sieg auf der Rennstrecke. Alles andere würde ja keinen Sinn machen. Denn sein überhaupt erster Sieg in der Formel 1, noch dazu mit Ferrari, war eben dort in diesem Jarama! Gleichzeitig erlöste der Gewinn Ferrari von einer sieglosen Durststrecke und obendrauf freuten sich alle, dass der Ferrari endlich funktionierte. Inspiration genug?! Jack Heuer pflichtete Niki Lauda bei, „dieser erste Ferrari-Sieg war sehr wichtig für uns".

Jack Heuer hatte immer ein gutes Gespür für Namen. Sie setzen sich zusammen aus schönen Städten, Rennstrecken oder hatten produktspezifische Attribute (Autavia: Auto-Aviation).

Die Heuer Jarama war nach Monza und Modena eine weitere Modellreihe mit direkter Verbindung zu Niki Lauda und Enzo Ferrari.

Im Laufe der Saison 1977 kühlte sich aber das Verhältnis zwischen Lauda und Ferrari merklich ab. So geschah es, dass der neue Weltmeister 1978 zu Braham Alfa Romeo dem Team seines Freundes Bernie Ecclestone wechselte. Enzo Ferrari reagierte sofort und verpflichtete den jungen Kanadier Gilles Villeneuve, der Lauda bereits bei den zwei letzten Rennen der Saison

Circuito del Jarama is a motorsport racetrack near San Sebastian los Reyes, 32 kilometer (19.9 miles) north of the Spanish capital of Madrid. The 3.4-kilometer (2.1 miles) circuit hosted the Spanish Formula 1 Grand Prix nine times between 1968 and 1981.

But how did the Heuer Jarama get its name?

After Niki Lauda's serious accident at the Nürburgring in 1976 and his resulting loss of the World Championship by one point to James Hunt, the new 1977 season went really well for the Austrian and the Scuderia. In the end, a fourth place in Watkins Glen (USA) was enough to secure Niki Lauda his second World Championship title in the premier class.

I asked Jack Heuer and Niki Lauda at a meeting in Vienna how the name Heuer Jarama came about: For Niki, it was logical that it must have been his special victory on the racetrack; anything else would be nonsense. After all, his first ever victory in Formula 1, and in a Ferrari, was at this very Jarama! At the same time, the win relieved Ferrari of a winless drought and, on top of that, everyone was happy that the Ferrari was finally working. Inspiration enough?! Jack Heuer agreed with Niki Lauda, 'this first Ferrari victory was very important for us'.

Jack Heuer always had a good nose for names. They were taken from beautiful cities, race tracks, product-specific attributes (Autavia: Auto-Aviation).

After Monza and Modena, the Heuer Jarama was another model series with a direct link to Niki Lauda and Enzo Ferrari.

During the 1977 season, however, the relationship between Lauda and Ferrari cooled noticeably. As a result, the new world champion switched to Braham Alfa Romeo, the team of his friend Bernie Ecclestone, in 1978. Enzo Ferrari reacted immediately and signed the young Canadian Gilles Villeneuve, who replaced Lauda in the Ferrari 312 T2 already for the last two races of the 1977 season.

1977 im Ferrari 312 T2 ersetzen sollte.
Die Heuer Jarama wird im deutschsprachigen Katalog von 1977 gleich mit drei verschiedenen Referenzen präsentiert. Die ersten zwei Modelle mit den Referenzen 110.245 und 110.225 hatten Stahlgehäuse mit schwarzem oder champagnerfarbigem Zifferblatt. Bei beiden Varianten wurde ein goldfarbenes, geriffeltes Bezel sowie Stahlband mit vergoldeten Mittelteilen verbaut.
Die dritte, Referenz 110.223, war eine reine Stahlvariante mit schwarzem Blatt, glatter schwarzer Lünette sowie Stahlband mit schwarzen Mittelteilen.
Eine vierte Version mit weißgoldener, geriffelter Lünette vervollständigte das Portfolio, allerdings ohne zugeordnete Referenznummer. Baugleich war sie mit der Referenz 110.223 inklusive der schwarzen Mittelglieder des Stahlbandes. Einziger Unterschied: Die schwarze Lünette wurde durch eine weißgoldene, geriffelte Lünette ersetzt. In der Recherche fällt auf, dass bereits ein Jahr nach Einführung der Heuer Jarama die 110.223 nicht mehr in den Preislisten auftaucht, sondern nur mehr die beiden Referenzen mit gelbgoldener Lünette. Die Uhren mit weißgoldener Lünette, die ich über die Jahre sah, hatten entweder Gehäuse mit der Referenz 110.233 oder 110.245, die ja sowieso ident waren.
Im Gegensatz zu anderen Heuer Bicolor-Modellen besaß die Jarama immer Drücker und Krone aus Stahl. Das kissenförmige Gehäuse hat oben einen oval gebürsteten Bereich, der das Bezel umrandet. Darauf folgen polierte Phasen, die seitlich Drücker und Krone integrieren. Referenz und Seriennummer finden sich bei 12- respektive 6 Uhr. Die Seriennummern sind im Bereich 356xxx angesiedelt. Der Rückdeckel hat einen Längsschliff und ist an der Innenseite signiert. Die Jarama-Gehäuse haben zwischen den Bandanstößen zwei unterschiedliche Pinbohrungen für Leder- oder Stahlband. Nach nur zwei Jahren Bauzeit stellte Heuer die Jarama Modellreihe ein. Im Heuer Katalog aus dem Jahr 1979 sowie in der dazugehörenden Preisliste fehlt die heute wie damals seltene Jarama. Sie ist mit ihrem besonderen Design und der Geschichte sicherlich ein Sammler-Tipp geworden.
Heuer zog sich mit Ende der Saison 1979 als Ferrari-Sponsor zurück. Eine innovative und erfolgreiche Kooperation zwischen Uhrenhersteller und Rennstall wurde im letzten Jahr nochmals mit einem Weltmeistertitel, gewonnen durch den Südafrikaner Jody Scheckter, gekrönt.

The Heuer Jarama was presented in the 1977 German-language product catalog with three different references. The first two models with the references 110.245 and 110.225 had steel cases with black or champagne-colored dials. Both versions had a gold-colored fluted bezel and a steel bracelet with gold-plated center parts.
The third, reference 110.223, was an all-steel version with a black dial, smooth black bezel, and steel bracelet with black center sections.
A fourth version with a white-gold-colored fluted bezel completed the portfolio, albeit without an assigned reference number. It was identical to reference 110.223, including the black center links of the steel bracelet. The only difference was that the black bezel was replaced by a white gold-colored fluted bezel. Research shows that just one year after the introduction of the Heuer Jarama, the 110.223 no longer appears in the price lists, but only the two references with a yellow-gold-colored bezel. The watches with white-gold-colored bezels that I have seen over the years had cases signed with reference 110.233 or 110.245, which were identical anyway.
In contrast to other Heuer bicolour models, the Jarama always had steel pushers and crown. The cushion-shaped case has an oval brushed area at the top that surrounds the bezel. This is followed by polished areas that integrate the side pushers and crown. The reference and serial number are located at 12 and 6 o'clock respectively. The serial numbers are in the 356xxx range. The back cover has a longitudinal cut and is signed on the inside. The Jarama cases have two different pin holes between the lugs for leather or steel straps. After only two years of production, Heuer discontinued the Jarama model series. The Heuer catalog from 1979 and the corresponding price list do not include the Jarama, which is as rare today as it was then. With its special design and history, it has certainly become a collector's tip.
Heuer withdrew as a Ferrari sponsor at the end of the 1979 season. An innovative and successful cooperation between the watch manufacturer and the racing team was crowned once again last year with a World Championship title, won by the South African Jody Scheckter.

TACHYMETER
PULSOMETER
JARAMA
HEUER
AUTOMATIC
CHRONOGRAPH
SWISS
NIKI
LAUDA
Marlboro
Marlboro

JARAMA

ref. 110.223 / 110.245

Jarama-Gehäuse aus Stahl mit verschraubtem Boden, oben mit Strichschliff und seitlich mit polierten Facetten. Eine weißgoldene, geriffelte Lünette ist am Gehäuse montiert. Das Stahlband mit schwarzen Mittelteilen wurde für die Jarama Linie designt.
Das Zifferblatt ist schwarz mit einem 30-Minuten- sowie 12-Stunden-Hilfszifferblatt und weißen rechteckigen Zeigern.
Abgerundet wird das Blatt von einem schwarzen Tachymeter-Ring mit roter Pulsometer -Skala zwischen 12 und 3 Uhr. Der Modelname „JARAMA“ steht über dem Heuer-Logo unter 12 Uhr sowie „AUTOMATIC CHRONOGRAPH“ über dem mit „SWISS“ signierten und weiß umrandeten Datumsfenster. Die länglichen Stundenindizes haben außen einen Leuchtpunkt. Keine Indizes findet man bei 3, 6 und 9 Uhr.
Weiße Stahlzeiger haben eine silberne Mitte und längliche Leuchtmasse-Einschlüsse. Der angespitzte weiße Chronographen-Sekundenzeiger berührt präzise die 1/5-Sekunden-Skala am äußeren Ende des Zifferblattes. Die signierte Krone sitzt links. Die Chronographen-Drücker sind rechts integriert und von hinten sichtbar. Der Rückdeckel zeigt einen Längsschliff.
Das Automatik-Uhrwerk ist ein Kaliber 12 und auf der Brücke mit „CAL.12“ signiert.

Steel Jarama screw-back case with brushed finish on top and polished facets on sides. A white-gold colored fluted bezel is mounted on the topcase. The steel band with black center links was designed for the Jarama Series.
The dial is black with subdials recording 30 minutes and 12 hours, and white hands. It is surrounded by a black tachymeter ring with an additional red printed pulsometer scale between 12 and 3 o'clock. Underneath 12 o'clock, "JARAMA" is above the Heuer logo. "AUTOMATIC CHRONOGRAPH" is above a "SWISS"-signed white-framed date window at 6 o'clock. Long hour markers have luminous dots on the outer track markers have been omitted at 3, 6, and 9 o'clock. steel hands are painted white with a silver center and luminous inserts. The tapering white chronograph hand reaches the 1/5-second track on the outer edge of the dial precisely. The crown on the left is signed Heuer. Chronograph pushers are integrated on the right of the case but fully visible from the back. The case back has longitudinal grinding.
The automatic movement is a Caliber 12 signed "CAL.12" on the bridge.

TACHYMETER
PULSOMETER
JARAMA
HEUER
AUTOMATIC
CHRONOGRAPH
SWISS

PULSOMETER
TACHYMETER
JARAMA
HEUER
AUTOMATIC
CHRONOGRAPH
SWISS
18

JARAMA

ref. 110.225

Jarama-Gehäuse aus Stahl mit verschraubtem Boden, oben mit Strichschliff und seitlich mit polierten Facetten. Eine gelbgoldene, geriffelte Lünette ist oben am Gehäuse montiert. Das Stahlband mit goldenen Mittelteilen wurde für die Jarama Linie designt.
Das Zifferblatt ist champagnerfarben mit einem 30-Minuten- sowie 12-Stunden-Hilfszifferblatt und schwarzen rechteckigen Zeigern.
Abgerundet wird das Blatt von einem champagnerfarbenen Tachymeter-Ring mit schwarzer Pulsometer-Skala zwischen 12 und 3 Uhr. Der Modelname „JARAMA" steht über dem Heuer Logo unter 12 Uhr sowie „AUTOMATIC CHRONOGRAPH" über dem mit „SWISS" signierten und schwarz umrandeten Datumsfenster. Die länglichen Stundenindizes haben außen einen Leuchtpunkt. Keine Indizes findet man bei 3, 6 und 9 Uhr.
Schwarze Stahlzeiger haben eine goldene Mitte und längliche Leuchtmasse-Einschlüsse. Der angespitzte schwarze Chronographen-Sekundenzeiger berührt präzise die 1/5-Sekunden-Skalaam äußeren Ende des Zifferblattes. Die signierte Krone sitzt links. Die Chronographen-Drücker sind rechts integriert und von hinten sichtbar. Der Rückdeckel zeigt einen Längsschliff. Das Automatik-Uhrwerk ist ein Kaliber 12 und auf der Brücke mit „CAL.12" signiert.

Steel Jarama screw-back case with brushed finish on top and polished facets on sides. A yellow-gold-colored fluted bezel is mounted on the top case. The steel band with golden center links was designed for the Jarama Series.
The champagne dial has subdials recording 30 minutes and 12 hours and black hands. It is surrounded by a champagne tachymeter ring with an additional black printed pulsometer between 12 and 3 o'clock. Underneath 12 o'clock, "JARAMA" is above the Heuer logo. "AUTOMATIC CHRONOGRAPH" is over the "SWISS"-signed black-framed date window at 6 o'clock. Long hour markers have luminous dots on the outer track; markers have been omitted at 3, 6, and 9 o'clock. The steel hands are painted black with a golden center and luminous inserts. The tapering black chronograph hand reaches the 1/5-second track on the outer edge of the dial precisely. The crown on the left is signed "HEUER." Chronograph pushers are integrated on the right of the case but fully visible from the back. The case back has longitudinal grinding. The automatic movement is a Caliber 12 signed "CAL.12" on the bridge.

TACHY
KENTUCKY
LUN 18
AUTOMATIC
SWISS
HEUER
TACHY
KENTUCKY
MON 18
AUTOMATIC
SWISS
HEUER
KENTUCKY
MON 25
AUTOMATIC
KENTUCKY
SUN 25
AUTOMATIC

KENTUCKY

Kentucky ist ein amerikanischer Bundesstaat im mittleren Osten. Aus dem indianischen kommend bedeutet Kentucky Feld oder Wiese.

Alljährlich, aber bereits seit 1875, findet in Louisville das Kentucky Derby statt. Das Rennen, auch „Run for the Roses" oder „Greatest Two Minutes" im Sport genannt, ist ein Galoppwettkampf für Vollblut-Pferde und eine der größten Sportveranstaltungen der USA. 2024 wurde eine Siegesprämie von 3,1 Mio Dollar an den Schnellsten über die Strecke von 1,25 Meilen ausbezahlt.

Die Technik der Zeitnehmung bei Pferderennen und der Formel 1 dürfte ähnlich sein. Heuer hatte beste Referenzen als „Timekeeper" von Ferrari im autoaffinen Amerika, somit ist es kaum verwunderlich, dass die Marketing-Strategen sich für eine neue Produktlinie namens Kentucky stark machten.

1977 war es dann soweit und Heuer lancierte die neue Modellreihe „Kentucky". Der Gehäusedurchmesser betrug 39 mm, gepaart mit einer neuen extravaganten Formensprache. Das Gehäusedesign war inspiriert von einem Hufeisen, welches das Zifferblatt umschloss und an der Oberseite kissenförmig auslief. Unten an der 6-Uhr-Position finden sich zwei Schlitzschrauben, die eine angewinkelte Fläche mit aufgesetztem reliefartigem Heuer-Logo begrenzen. Der Kentucky-Automatik-Chronograph hat als einzige Uhr im Produktkatalog von 1977 kein Heuer-Logo auf dem Zifferblatt!

Zwei am Gehäuse angebrachte Stifte dienten wie eine Schiene als Befestigung für das speziell entworfene Stahlband mit Heuer signierter und verriegelbarer Sicherheitsschließe.

Angetrieben wurde die Kentucky Referenz 750.703/5 durch ein Valjoux 7750-Automatik Werk mit Chronographen-Funktion (750 stand für Valjoux 7750, 7 für Modellreihe Kentucky, 3/5 für Stahl oder Stahlgold). Eine gerahmte Anzeige für Wochentag und Datum findet sich bei 3 Uhr. Darüber steht in Großbuchstaben der Modell-Name „KENTUCKY", darunter „AUTO-

Kentucky is a state in the south-east of the USA. Its name comes from the native American word for field or meadow.

The Kentucky Derby has been held in Louisville every year since 1875. The race, also known as the "Run for the Roses" or "Greatest Two Minutes" in sport, is a flat race for thoroughbred horses and one of the biggest sporting events in the USA. In 2024, a prize of 3.1 million dollars was paid out to the fastest horse over the 1.25-mile course.

The timekeeping technology used in horse racing and Formula 1 is likely to be similar. Heuer had the best references as Ferrari's "timekeeper" in the car-loving USA, so it is hardly surprising that the marketing strategists were in favor of a new product line called Kentucky. In 1977, the time had come and Heuer launched the new Kentucky model series. The 39-mm (1.54-in) diameter case was paired with a new, striking design language. The case design was inspired by a horseshoe, which surrounded the dial and flared out in a cushion shape at the top. At the bottom, at the 6 o'clock position, are two slotted screws that border a beveled surface with a relief-like Heuer logo. The Kentucky automatic chronograph is the only watch in the 1977 product catalog without a Heuer logo on the dial!

Two pins attached to the case formed a bar for attaching the specially designed steel bracelet with a Heuer-signed lockable safety clasp.

The Kentucky with reference 750.703/5 is powered by a Valjoux 7750 automatic movement with chronograph function (750 stands for Valjoux 7750, 7 for Kentucky model series, 3/5 for steel or steel/gold). A framed display for the day of the week and date can be found at 3 o'clock. Above the day-date frame, "KENTUCKY" is shown in capital letters above "AUTOMATIC" and "SWISS". The permanent running second at 9 o'clock and the 30-minute and 12-hour subdials between 12 and 6 o'clock are predetermined by the movement. The pushers and crown are classically positioned on the right.

HEUER-Werbung aus dem Jahr 1978.

HEUER advertisment from 1978.

MATIC" und verkleinert „SWISS". Durch das Uhrwerk vorgegeben sind die permanente Sekunde bei 9 Uhr und in einer Achse zwischen 12 und 6 Uhr sind die 30-Minuten- und 12-Stunden-Hilfszifferblätter positioniert. Drücker und Krone sind klassisch auf der rechten Seite angeordnet. Der polierte Rückdeckel hat Öffnungsnuten an der Außenseite und mittig eine gebürstete Fläche mit Serien- und Referenznummer.

Es existieren vier unterschiedliche Kentucky Automatik-Modelle, alle basierend auf dem Valjoux 7750. Die Kentucky in Stahl gab es mit blauem (B) oder schwarzem (N) Zifferblatt, sowie in Stahlgold mit schiefergrauem (G) und schwarzem (N) Zifferblatt. Bei den Stahlgold-Uhren waren das Heuer-Relief, Schrauben, Krone und Drücker vergoldet. Das Stahlband war bei allen vier Varianten ident ohne jegliche Gold-Applikationen.

Die „Kentucky" verschwand 1979 nach zwei Jahren wieder aus dem Heuer-Katalog. Die blaue Zifferblattvariante Referenz 750.503B ist sehr selten, aber richtig „cool" und dem damaligen Zeitgeist entsprechend ist die Referenz 750.705G mit schiefer-grauem Blatt und kontrastiven Goldapplikationen.

The polished back cover has opening grooves on the outside and a brushed surface with serial and reference numbers in the center.

There were four different Kentucky automatic models, all based on Valjoux 7750 movements, available in steel with a blue (B) or black (N) dial and in steel gold with a slate (G) and black (N) dial. On the steel gold watches, the Heuer relief, screws, crown, and pushers were gold-plated. The steel bracelet without gold elements was identical on all four.

The Kentucky disappeared from the Heuer catalogue after two years in 1979. The blue dial variant reference 750.503B is very rare, but the reference 750.705G with slate dial and contrasting gold elements is very cool and in keeping with the spirit of the time.

Heuer stoppt die Zeit für Anästhesisten, Arbeitsvorbereiter, Ballonflieger, Basketballer, Betreuer, Bobsleigher, Boxenmechaniker, Boxer, Cardiologen, Chemiker, Cost-Controller, Cross-Country-Cracks, Disc-Jockeys, Economy-Rallye-Cracks, Eishockey-Cracks, Eisschnelläufer, Elektronentechniker, Fallschirmspringer, Fechter, Fernfahrer, Fernschreiber, Fernsehsprecher, Flugkapitäne, Fotografen, Fußballer, Grand-Prix-Piloten, Gynäkologen, Handballer, High-Fidelity-Fans, Hochseefischer, IFR-Piloten, Ingenieure, Jockeys, Judokas, Kameraleute, Kanuten, Kunstflieger, Kunstturner, Laboranten, Langläufer, Leichtathleten, Lotsen, Meteorologen, Military-Reiter, Modellflieger, Motocross-Fahrer, Motorbootrennfahrer, Motorjournalisten, Navigatoren, Physiker, Piloten, Polo-Spieler, Psychologen, Radfahrer, Rallye-Fahrer, Redner, Regatta-Segler, Regisseure, Rennläufer, Rennstallbesitzer, Ringer, Rodeo-Reiter, Ruderer, Rugby-Spieler, Rundfunksprecher, Schiedsrichter, Schiffsoffiziere, Schützen, Schwimmer, Sendeleiter, Ski-Rennläufer, Skispringer, Sprungreiter, Strömungstechniker, Taucher, Telefonisten, Testfahrer, Tontechniker, Trainer, Trial-Cracks, Unterwasserjäger, Yacht- Cracks, Wasserballer, Wasserskisportler, Zeitfahrer und Zeppelinkapitäne.

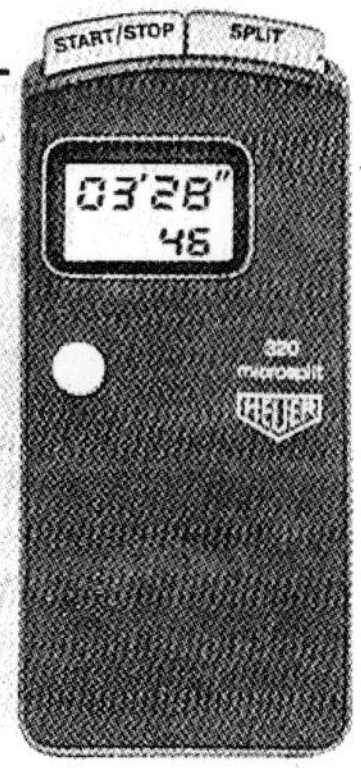

Microsplit 320. Elektronische Quarz-Stoppuhr mit LCD-Anzeige. 1/100 Sekunden-Messer mit Split-Betrieb. Meßbereich bis 59 Min. 59,99 Sek. Eingebaute Tragkordel.

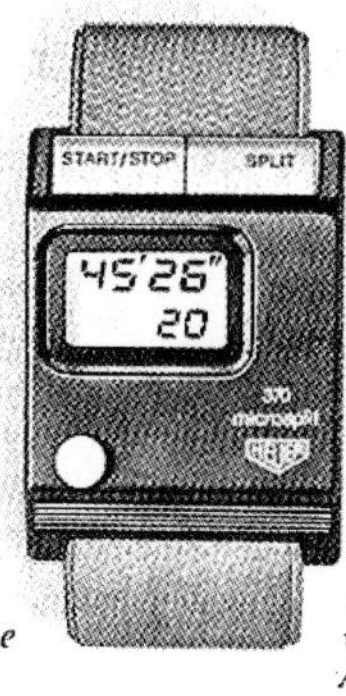

Microsplit 370. Elektronische Quarz-Stoppuhr (techn. Daten wie Modell 320) Wassergeschütztes Gehäuse mit elastischem, verstellbarem Armband.

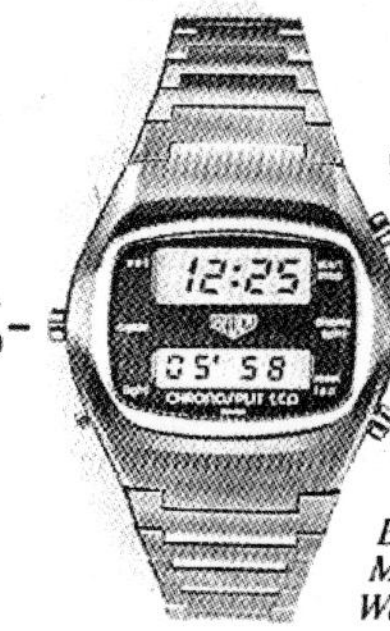

Chronosplit LCD. Der erste und einzige Quarz-Chronograph der Welt mit doppelter Digitalanzeige. Zwischenzeitspeicher. Edelstahlgehäuse. Mineralglas. Wasserdicht.

Kentucky. Automatischer Chronograph mit 12-Stunden-Zähler und Datum. Edelstahlgehäuse. Wasserdicht. Leuchtzifferblatt. Mineralglas. Tachymeter- Einteilung.

HEUER

Bezugsquellennachweis: Chronoswiss, A-1014 Wien, Tuchlauben 8, Telefon (0 22 2) 63 07 16

GGK

KENTUCKY

ref. 750.703 B

Stahlgehäuse mit verschraubtem Boden und Strichschliff auf der Gehäuseoberseite und den Flanken. Am Gehäuse neben 6 Uhr sitzen zwei stählerne Schlitzschrauben, darunter eine angewinkelte Fläche mit aufgesetztem reliefartigem Heuer-Logo.
Hier eine Version mit blauem Zifferblatt, und drei Hilfszifferblättern, diese zeigen eine permanente Sekunde bei 9 Uhr, eine 30-Minuten- bzw. 12-Stunden-Einteilung in einer Achse von 12 nach 6 Uhr an, umgeben von einem blauen Tachymeter-Ring mit weißen Zahlen. Der Modellname „KENTUCKY" steht weiß gedruckt über dem metallgerahmten Tag-Datumsfenster bei 3 Uhr, darunter „AUTOMATIC" und „SWISS". Die Indizes aus Metall mit kleinen Leuchtpunkten an der Innenseite fehlen bei 12, 3, 6 und 9 Uhr.
Das rechteckige Minuten-Stunden-Zeigerspiel ist weiß umrandet, innen gebürstet, mit Leuchtmaterial-Einlage; der weiße Chronographen-Zeiger berührt mit seiner Spitze die Minutenskala am äußeren Rand des Zifferblattes.
Krone und Drücker sind rechts im Gehäuse integriert. Der Rückdeckel ist seitlich angewinkelt und poliert, mittig befindet sich eine gebürstete kreisrunde Fläche mit Referenz- und Seriennummer. Bei dieser Uhr nicht sichtbar aufgrund des Heuer-NOS (New Old Stock)-Stickers.
Das Uhrwerk ist ein Valjoux 7750, bei der Unruhe signiert sowie mit Heuer-Logo auf dem Automatikrotor.

Steel screw-back case with line grinding on the top and sides. There are two steel slotted screws on the case next to 6 o'clock, below is an angled surface with Heuer logo relief.
Blue dial version with three dials showing a running second at 9 o'clock and a 30-minute and 12-hour division between 12 and 6 o'clock: surrounded by a blue tachymeter ring with white numerals.
The "KENTUCKY" model name is printed in white above the metal-framed day-date window at 3 o'clock, with "AUTOMATIC" and "SWISS" below it. The metal markers with small luminous dots on the inside are omitted at 12, 3, 6, and 9 o'clock.
The rectangular minute/hour hand is framed in white and brushed on the inside, with luminous inlay; the tip of the white chronograph hand touches the minute scale on the outer edge of the dial.
The crown and pushers are integrated into the right of the case. The back cover is beveled and polished; in the center is a brushed circular surface with reference and serial numbers. Not visible on this example due to an applied NOS (New Old Stock) Heuer sticker.
The movement is a Valjoux 7750, signed on the balance wheel and with the Heuer logo on the automatic rotor.

TACHY
KENTUCKY
LUN 18
AUTOMATIC
SWISS
HEUER

KENTUCKY

ref. 750.705 G

Stahlgehäuse mit verschraubtem Boden und Strichschliff auf der Gehäuseoberseite und den Flanken. Am Gehäuse neben 6 Uhr sitzen zwei vergoldete Schlitzschrauben, darunter eine Fläche mit goldfarbenem reliefartigem Heuer-Logo.
Hier eine Version mit schiefergrauem Zifferblatt und drei Hilfszifferblättern, diese zeigen eine permanente Sekunde bei 9 Uhr sowie 30-Minuten- bzw. 12-Stunden-Einteilung in einer Achse von 12 nach 6 Uhr an, umgeben von einem goldfarbenen Tachymeter-Ring mit schwarzen Zahlen. Der Modellname KENTUCKY steht schwarz gedruckt über dem gerahmten Tag-Datumsfenster bei 3 Uhr, darunter AUTOMATIC und SWISS. Die Indizes aus Metall mit Leuchtpunkten an der Innenseite fehlen bei 12, 3, 6 und 9 Uhr.
Das rechteckige Minuten-Stunden-Zeigerspiel ist schwarz umrandet, innen gebürstet mit Leuchtmaterial-Einlage, der schwarze Chronographen-Zeiger berührt mit seiner Spitze die Minutenskala am äußeren Rand des Zifferblattes.
Die vergoldete Krone und Drücker sind rechts im Gehäuse integriert. Der Rückdeckel ist seitlich angewinkelt und poliert, mittig befindet sich eine gebürstete, kreisrunde Fläche mit Referenz- und Seriennummer.
Das Uhrwerk ist ein Valjoux 7750, bei der Unruhe signiert, sowie mit Heuer-Logo auf dem Automatikrotor.

Steel screw-back case with line grinding on the top and sides. There are two gold-plated slotted screws on the case next to 6 o'clock; below is an angled surface with a gold-colored Heuer logo relief.
Slate-gray dial version with three subdials showing a running second at 9 o'clock and 30-minute and 12-hour sub dials on an axis at 12 to 6 o'clock, surrounded by a gold-colored tachymeter ring with black numerals.
The "KENTUCKY" model name is printed in black above the framed day-date window at 3 o'clock, with "AUTOMATIC" and "SWISS" below it. The metal indexes with luminous dots on the inside are missing at 12, 3, 6, and 9 o'clock.
The rectangular minute/hour hand is edged in black, brushed on the inside with a luminous inlay; the tip of the black chronograph hand touches the minute scale on the outer edge of the dial.
The gold-plated crown and pushers are integrated into the right of the case. The back cover is beveled and polished, with a brushed circular surface in the center bearing the reference and serial numbers.
The movement is a Valjoux 7750, signed on the balance wheel and with the Heuer logo on the automatic rotor.

TACHY
KENTUCKY
MON 18
AUTOMATIC
SWISS
HEUER

HEUER

MILITARY

Die Uhrenmanufaktur Heuer in Biel zählte zu den führenden Produzenten für Stoppuhren und begann in den frühen 1960er Jahren das Chronographen-Portfolio auszubauen.
Mit Einführung neuer Flugzeug-Generationen, wie der Boeing 707 oder Douglas DC-8, angetrieben durch vier Strahltriebwerke, verkürzten sich die Flugzeiten bei gleichzeitig gestiegenen Kapazitäten. Die internationale Luftfahrt entwickelte sich rasant von der Nische zum Produktschlager.
Flugzeugborduhren in Cockpits oder für die Navigation adaptierte Chronographen durften im wachsenden Heuer-Portfolio nicht fehlen.
Mitten im Kalten Krieg, geprägt durch machtpolitische Rivalität zwischen Ost und West, wurde in den 60er Jahren massiv aufgerüstet. Die Bundeswehr bestellte bei Lockheed über 900 Starfighter-Kampfjets, die sukzessive nach Deutschland überstellt wurden, daher stieg der Bedarf an Pilotenuhren.
Zu dieser Zeit fusionierte Heuer 1964 in der Schweiz mit dem früheren Konkurrenten Leonidas und erwarb Spezialwissen im Bereich Militäruhren sowie wichtige zusätzliche Produktionskapazitäten. Sich Lieferant des Militärs nennen zu dürfen, war für einen Uhrenlieferanten ein besonderes Qualitätsmerkmal mit hohem Prestige-Faktor.
Die Abteilung für Beschaffung in Koblenz hatte verschiedene Produzenten eingeladen, eine neue Generation von Dienstuhren zu bemustern. Die Ausschreibung der deutschen Bundeswehr stellte folgende Anforderung:
Chronograph mit Doppelstoppeinrichtung sowie Ankerwerk aus Schweizer Produktion; stoßsicher mit 17-Rubin-Lager; Sekundenstoppmechanismus (durch Ziehen der Krone wird eine kleine mitlaufende Sekunde bei 9 Uhr arretiert, dadurch wird das Synchronisieren mehrerer Uhren erleichtert).
Sowohl Zeiger als auch arabische Ziffern mussten flächig mit Tritium belegt sein, um in der Dunkelheit zu leuchten.
Für die zwei Hilfszifferblätter schrieb der Bund unterschiedlich geformte Zeiger vor. Charakteristisch und gut ablesbar ist der Chronographen-Zeiger mit zum Dreieck geformter Spitze.

Based in Biel, Switzerland, Heuer was one of the leading producers of stopwatches and began to expand its chronograph portfolio in the early 1960s.
The introduction of new generations of aircraft, such as the Boeing 707 and Douglas DC-8, powered by four jet engines, shortened flight times while increasing passenger capacity. International aviation developed rapidly from niche to best-seller.
Aircraft clocks in cockpits or chronographs adapted for navigation simply had to be part of the growing Heuer portfolio.
In the middle of the Cold War and its by power-political rivalry between East and West, the 1960s saw rearmament on a massive scale.
The German armed forces ordered over 900 Starfighter fighter jets from Lockheed, which were successively transferred to Germany, thus increasing the demand for pilot watches.
At that time in Switzerland, Heuer merged with its former competitor Leonidas in 1964 and acquired specialized knowledge in the military watch sector as well as important additional production capacities.
The right to the title of "supplier to the military" was a special quality feature with a high prestige factor for a watch manufacturer!
The Bundeswehr procurement department in Koblenz invited various manufacturers to submit samples for a new generation of service watches.
The tender had the following requirements:
Chronograph with double stop device and Swiss-made lever movement; shock-resistant with 17 jewel movement; "hacking mechanism" (by pulling the crown, a small running second at 9 o'clock is locked, making it easier to synchronize several watches).
Both the hands and the Arabic numerals had to be coated with tritium to glow in the dark.
For the two subdials, the tender specified differently shaped hands. The chronograph hand with triangular tip is characteristic and easy to read.
Heuer designed the new reference 1550 SG with a two-part stainless-steel case. The case back was shaped like a cage and connected to the upper part by four screws. The acrylic glass with sealing ring, together with the sealed crown and pushers, prevents moisture from entering.

Heuer bemusterte mit der neuen Referenz 1550 SG ein zweiteiliges Gehäuse aus Edelstahl. Der Gehäuseboden war wie ein Käfig geformt und mittels vier Schrauben mit dem Oberteil verbunden. Das Acrylglas mit Dichtungsring verhindert im Zusammenspiel mit abgedichteter Krone und Drückern den Feuchtigkeitseintritt.
Die Gehäusekanten, so forderte die Bundeswehr, sollten bei der neuen Heuer runder sein, damit keine Verletzungsgefahr bestand. Auch wurde das Gehäuseoberteil mattiert, um Reflexionen zu minimieren.
Eine breite Lünette mit Minuten-Skala lässt sich in beide Richtungen drehen.
Das Herzstück bildet ein Kaliber Valjoux 230, eine Abwandlung des Valjoux 23, mit für militärische Einsätze wichtiger „Flyback"-Funktion.
Zur Synchronisierung der Uhren betätigt man den Drücker bei 4 Uhr, der Chronograph wird zurückgestellt und sofort bei Loslassen neu gestartet.
Das Valjoux 230 Werk hatte eine garantierte Gangreserve von mindestens 36 Stunden.
Die Ganggenauigkeit wurde in den Spezifikationen der Bundeswehr genau definiert, bei einer maximalen Abweichung von +/– 10 s pro Tag oder 24 Stunden bei einer Temperatur von 20 °C mit eingeschaltetem Chronographen.
Die Uhren wurden bei der Bundeswehr in zwei Positionen getestet: Zuerst horizontal mit dem Zifferblatt oben, dann vertikal mit der Aufzugskrone unten.
Mit der finalen Erteilung des Auftrages begann Heuer umgehend die gewünschten Adaptionen der neuen Heuer Referenz 1550 SG für die Serienproduktion umzusetzen. Die Bundeswehr war unter großem Zeitdruck.
Heuer ergänzte sogar 255 Leonidas-Chronographen aus Lagerbeständen, die mit Preisabschlag unverzüglich nach Deutschland geliefert wurden.
Das Gehäuse der Leonidas CP-1, die bereits 1964 für die italienischen Aeronautica Militare Italiana (AMI) produziert wurde, unterschied sich mit 39 mm Durchmesser maßgeblich vom direkten Nachfolger CP-2 mit 43 mm. Die deutsche Leonidas sowie CP-1 und CP-2 wurden durch das Valjoux 222 (entspricht dem Valjoux 22 mit der Zusatzfunktion Flyback – die letzte „2" in der Werknummerierung steht für Flyback) angetrieben.
Ab 1967 wurden nur mehr Chronographen mit Heuer-Logo auf dem Zifferblatt und einem Uhrwerk Valjoux 230 geliefert.
Mit wachsendem Bestand begann die Verteilung an andere Bereiche innerhalb der Bundeswehr. Neben der Luftwaffe bekamen auch Marineflieger, nautische Offiziere, U-Boot-Fahrer, Fernspäher und andere Spezialeinheiten den Heuer-Chronographen.
Fortan gehörten Heuer-Uhren zur Ausstattung von Schiffen. Diese Uhren blieben immer an Bord, wurden bei Bedarf ausgegeben und nach dem Einsatz wieder zurückgegeben.
Ein ehemaliger Offizier erzählte mir, dass ein deutsches Forschungsschiff mit dem klingenden Namen „Planet" noch immer zwei Heuer-Chronographen an Bord habe

The Bundeswehr required the case edges of the new Heuer to be rounded to prevent the risk of injury. The upper part of the case was also matt-finished to minimize reflections.
A wide bezel with a minute scale can be turned in both directions.
At its heart is a Valjoux 230 caliber, a modification of the Valjoux 23 with a flyback function that is important for military operations.
Watches are synchronized by pressing the pusher at 4 o'clock; the chronograph resets and is restarted as soon as it is released.
The Valjoux 230 movement had a guaranteed power reserve of at least 36 hours.
The accuracy was defined in the Bundeswehr specifications at a temperature of 20 °C with the chronograph switched on and a maximum deviation of +/-10 s per day or 24 hours.
The watches were tested by the Bundeswehr in two positions: first horizontally with the dial at the top, then vertically with the winding crown at the bottom.
When the order was finally placed, Heuer immediately began to carry out the desired adaptations of the new Heuer Reference 1550 SG for series production. The Bundeswehr was under great time pressure.
Heuer even completed 255 Leonidas chronographs from stock, which were immediately delivered to Germany at a discount.
The case of the Leonidas CP-1, which was produced for the Italian Aeronautica Militare Italiana (AMI) as early as 1964, differed significantly from its direct successor, the CP-2, with a diameter of 39 mm (1.54 in) instead of 43 mm(1.69 in).
The German Leonidas and the CP-1 and CP-2 were powered by the Valjoux 222 (corresponding to the Valjoux 22 with additional flyback function, denoted by the last 2 in the movement numbering).
From 1967, only chronographs with the Heuer logo on the dial and Valjoux 230 movement were supplied.
As stocks were built up, distribution to other areas within the Bundeswehr began. In addition to the air force, navy pilots, nautical officers, submariners, scouts, and other special units also received Heuer chronographs.
From then on, Heuer watches were part of ship equipment. These watches remained permanently on board, issued when required and returned after use.
A former officer told me that a German research vessel with the illustrious name "Planet" still has two Heuer chronographs on board, ready for use 50 years after first being put into service!
The large dial of the Heuer watches, over three centimeters in diameter, has excellent readability and the robustness of the construction extended their official service life.
Although new chronographs were purchased in the early 1980s, many military personnel are said to have refused to "decommission" their Heuer watches because the new ones, for example, no longer had the flyback

HEUER
STERNZEIT REGULIERT

– 50 Jahre nach der ersten Indienststellung noch immer bereit zum Einsatz!

Mit einem Durchmesser von über drei Zentimetern lässt sich das Zifferblatt ausgezeichnet ablesen und die robuste Konstruktion verlängerte die offizielle Dienstzeit der Heuer-Uhren.

Obwohl Anfang der 80er Jahre neue Chronographen angeschafft wurden, sollen viele Militärangehörige die „Außerdienststellung“ ihrer Heuer-Uhren verweigert haben, da die „neuen“ zum Beispiel nicht mehr über die „Flyback“-Funktion verfügten!

Die Ausmusterung der Heuer über die VEBEG (Verwertungsunternehmen der Bundesrepublik Deutschland für ausgemustertes Eigentum) begann in den 90er Jahren. Gemäß der Tradition fanden gut erhaltene Exemplare schon zuvor bei internen Personalkäufen rasanten Absatz und kamen infolgedessen im Originalzustand auf den Uhrenmarkt.

Andere Uhren – manche sagen die besseren – hatten nicht so viel Glück und wurden durch den Drillbohrer zerstört, da es sich ja offiziell um Kriegsgerät handelte.

Ordnungsgemäß ausgemustert und freigegeben für den sekundären Markt wurden zumeist Uhren in stark gebrauchtem, sehr schlechtem oder defektem Zustand, die nicht mehr für den Einsatz taugten – in Säcken zu 20 bis 70 Stück.

Ein Militaria-Händler, der damals aktiv diese ausgemusterten Uhren aufkaufte, erinnert sich an eine Gesamtstückzahl von grob geschätzt 1200 bis 1500 Uhren, die so von der Bundeswehr verkauft wurden. Zweithändler kauften diese Uhren auf, überarbeiteten sie, statteten sie mit anderen Zifferblättern aus und boten sie über Zeitungsannoncen an.

Während dieser Zeit verloren viele Heuer-Bund-Exemplare ihre Originalität. Die damals herrschende mangelnde Nachfrage ist heute lange vorbei. Denn Militär-Uhren mit Geschichte und tollem Design sind mittlerweile sehr begehrte Sammlerstücke mit positiver Wertentwicklung.

Das Zifferblatt gab es in vier Varianten:

1. Groß 3H, mit kleinem T über 6 Uhr
2. Mittel 3H, mit kleinem T über 6 Uhr
3. Klein 3H, mit kleinem T über 6 Uhr
4. Kein 3H, mit kleinem T über 6 Uhr

Das kleine T direkt über 6 Uhr ist ein wichtiger Indikator für die Identifizierung originaler Heuer-Zifferblätter. Wie das 3H steht auch T für Tritium.

Die Minutenskala am Zifferblattrand ist klar strukturiert und gut ablesbar. Die Beschichtung der Ziffern mit Tritium variiert je nach Einsatzgebiet in Farbe und Dicke. Das Heuer-Logo stand unter 12 Uhr und sollte nicht breiter als 5 mm sein.

Die Rückseite mit kreisförmigem Bürstenschliff wurde mit „Bund“ oder ausgeschrieben „Bundeswehr“ über der 13-stelligen NATO-Nummer graviert; Marineuhren waren oft nur mit der NATO-Nummer gekennzeichnet.

Neben den zuvor beschriebenen Dienstuhren gab es eine nach Sternzeit regulierte Variante – ein Messin-

function!

The decommissioning of the Heuer via the VEBEG (the Federal Republic of Germany's surplus sales company for decommissioned property) began in the 1990s. According to tradition, well-preserved examples had already sold rapidly in the Bundeswehr's internal personnel purchases and some were subsequently placed on the watch market in their original condition.

Other watches—as some say, the better ones—were not so lucky and were destroyed by drilling, as they were officially war equipment.

The watches that were properly decommissioned and released for the secondary market were mostly in very used, very poor, or even defective condition that were no longer suitable for use. They were sold in bags of 20 to 70 pieces.

A militaria dealer who actively bought up those decommissioned watches at the time remembers a total of roughly 1200 to 1500 watches being sold by the Bundeswehr in this way.

These watches were bought by second-hand dealers who refurbished them, fitted them with off-the-shelf dials, and offered them for sale in newspaper adverts.

During this time, many Heuer Bund models lost their originality. This lack of demand has long since passed. Military watches with history and great design have become highly sought-after collector's items and increased in value.

There are four different versions of the dial:

1. Large 3H, with a small T above the 6 o'clock position
2. Medium 3H, with a small T above the 6 o'clock position
3. Small 3H, with small T above the 6 o'clock position
4. No 3H, with small T above the 6 o'clock position

The small T directly above the 6 o'clock position is an important indicator for identifying original Heuer dials! Like T, 3H stands for tritium.

The minute scale on the edge of the dial is clearly designed and easy to read. The color and thickness of the tritium applied to the numerals varies depending on the area of application. The Heuer logo was positioned below 12 o'clock and should not exceed 5 mm (0.2 in) in length.

The back with circular brushed finish was engraved with "Bund" or "Bundeswehr" in full above the 13-digit NATO number; navy watches were often only labelled with the NATO number.

In addition to the service watches discussed above, there was also a sidereal time version, a special kind of measuring instrument. As the earth rotates once around its own axis within 23 hours, 56 minutes and 3.5 seconds according to sidereal time, sidereal time is required to determine coordinates using celestial bodies on the earth. Adjusted to sidereal time, the clock has a cycle of 3 minutes and 56.5 seconds per day. To avoid confusion, Heuer printed the words "STERNZEIT REGULIERT" (regulated sidereal time) above the 6 o'clock position on the dials.

The reference number was also changed from 1550 SG

strument der besonderen Art. Da sich die Erde nach der Sternzeit innerhalb von 23 Stunden, 56 Minuten und 3,5 Sekunden einmal um die eigene Achse dreht, bedarf es zur Bestimmung von Koordinaten mittels Gestirnen der Sternzeit. Nach Sternzeit reguliert weist die Uhr einen Vorgang von 3 Minuten und 56,5 Sekunden pro Tag auf. Um Verwechslungen vorzubeugen, bedruckte Heuer die Zifferblätter über 6 Uhr mit „STERNZEIT REGULIERT“.

Auch die Referenznummer wurde von 1550 SG in 1551 SGSZ (wobei SZ für Sternzeit steht) geändert.

Die nach Sternzeit regulierte Heuer wurde ausgegeben in einer schwarzen Ledermappe mit Reißverschluss, deren Front- und Rückseite mit astronomischen Anleitungen, der Zusatzausstattung, der Versorgungsnummer 6675-12-308-6782 sowie mit „Bund“ beschriftet waren. Die Ledermappe beinhaltete eine drehbare Sternkarte des nördlichen Sternenhimmels mit durchsichtigem Deckblatt und verstellbarem Zeiger. Diese Sternkarte diente dazu, Sterne und Sternbilder zu finden, Positionen von Planeten sowie die Sternzeit und den Stundenwinkel zu ermitteln. Zusätzlich findet man einen Reduktionsmaßstab für geographische Breite und Länge. Ein Ersatzband, eine Brille mit Etui sowie ein astronomisches Jahrbuch für die Truppenvermessung komplettieren das Set. Von den 50 ursprünglichen Sets sollen nur 36 Stück geliefert worden sein, wodurch die Referenz 1551SGSZ neben den Marine-Varianten sicher zun den seltensten Heuer-Modellen gehört.

to 1551 SGSZ (where SZ stands for Sternzeit).

The Heuer Sternzeit Reguliert was issued with a black leather folder with a zip fastener, which was marked on the front and back with astronomical direction, additional features, the supply number 6675-12-308-6782 and Bund.

The leather folder contained a rotating cosmos star chart of the northern night sky with a translucent cover sheet and adjustable pointer. This star chart was used to locate the stars and constellations, to determine the positions of the planets as well as the sidereal time and the hour angle. There is also a reduction scale for latitude and longitude. A spare strap, glasses with case and an astronomical yearbook for surveying troops complete the set. Only 36 of the 50 original sets are said to have been delivered.

This certainly makes the reference 1551SGSZ the rarest Heuer alongside the marine variants.

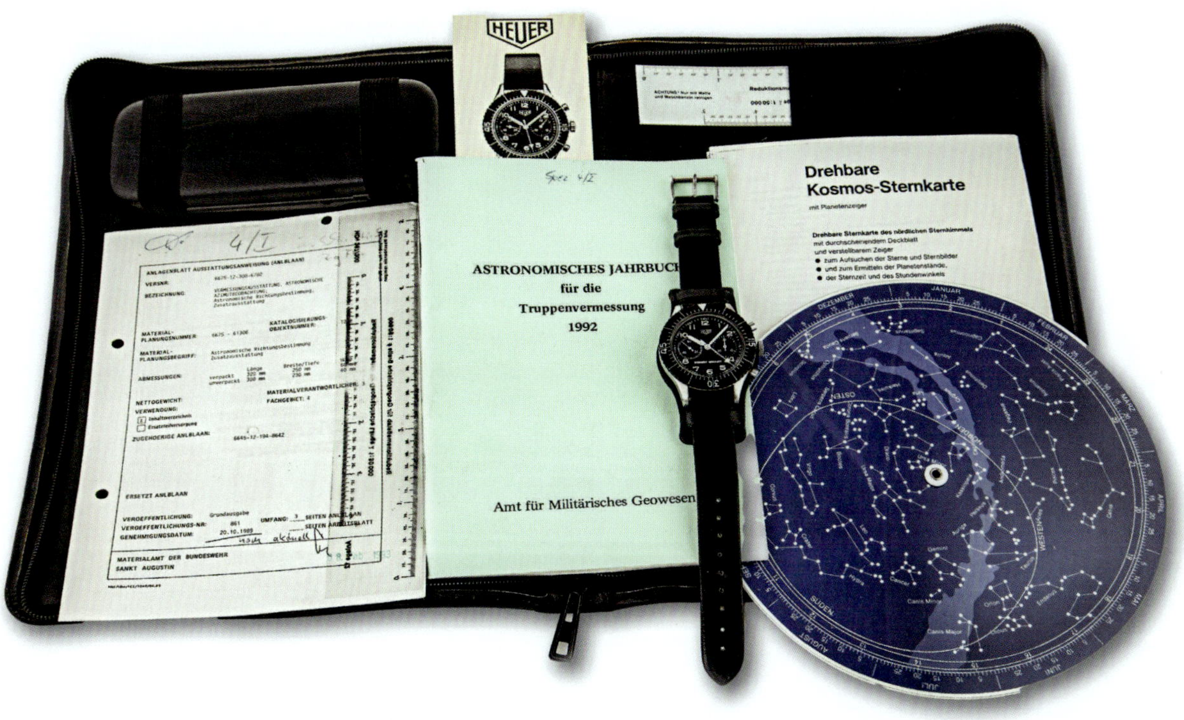

A.M.I.
CRONOMETRO
TIPO CP-1
M.M. 270280
1964

LEONIDAS

TIPO CP-1 A.M.I.

Stahlgehäuse, verschraubter Boden mit 4 Schrauben. Das mattschwarze Blatt hat arabische Ziffern von 1 bis 12. Diese sind mit Leuchtmasse belegt. Eine Skala am Blattrand wechselt zwischen Strichen für jede Minute und arabischen Zahlen bei 5, 10, 15, ... bis zur 60.
Die beiden Hilfszifferblätter haben konzentrische Kreise. Die Sekunden-Skala mit einer Stricheinteilung von 1 bis 60 (beziffert 20, 40, 60) befindet sich auf neun Uhr. Gegenüber ist das 30-Minuten-Hilfszifferblatt, beziffert mit 10, 20, 30. Alle Skalen sind weiß.
Die weißen Zeiger sind mit Tritium beschichtet. Die Spitze des Chronografen-Zeigers ist dreieckig, die Zeiger der zwei Hilfszifferblätter sind unterschiedlich geformt. Anstelle der üblichen 60 hat die schwarze Lünette eine silberne Dreiecksmarkierung, ansonsten Minutenstriche von 1 bis 59. Die 5-Minuten-Markierungen sind über die ganze Breite der Lünette gezogen und beziffert mit 15, 30, 45.
Das Uhrglas aus Kunststoff ist mit einer Dichtung versehen.
Das Gehäuse hat scharfe Kanten mit einer Bandanstoßbreite von 20 mm. Der Gehäusedeckel zeigt einen kreisförmigen Schliff.
Auf der Rückseite graviert ist „A.M.I" über „CRONOMETRO", darunter „TIPO CP-1". Wieder darunter findet man die Identifikationsnummer der Uhr M.M. 270280 und die Jahreszahl der Indienststellung 1964.
Das Uhrwerk ist mit Valjoux signiert, die Brücke ist unsigniert.
Ein mattiertes feinmaschiges Stahlzugband mit offener Gliederstruktur ist durch eine innenlaufende Schnur in der Länge justierbar.

Steel screw-back case with 4 screws. The matt black dial has Arabic numerals from 1 to 12, which are coated with luminous material.
A scale on the edge of the dial alternates between graduations for each minute and Arabic numerals at 5, 10, 15, to 60.
The two subdials have concentric circles. The seconds scale with graduations from 1 to 60 (numbered 20, 40, 60) is located at the nine o'clock position. Opposite is the 30-minute counter, numbered 10, 20, 30. All scales are printed in white.
The white hands are coated with tritium. The chronograph hand has a delta-shaped tip. The hands of the two subdials are differently shaped.
The black bezel has a silver triangular marking, otherwise 60 minute markers from one to 59. The five-minute markers are numbered 15, 30, 45 across the entire width of the bezel.
The plastic crystal is fitted with a seal.
The case has sharp edges with a lug width of 20 mm (0.79 in). Concentric circular grinding is visible on the case lid.
Engraved on the back is "A.M.I." above "CRONOMETRO," and below "TIPO CP-1"; below this is the identification number of the watch, M.M. 270280, and the year of commissioning, 1964.
The Valjoux signed movement has an unsigned bridge.
A matt-finish fine-meshed steel bracelet with an open link structure can be adjusted in length by an internal cord.

HEUER

ref. 1550 SG Large 3H

Stahlgehäuse, verschraubter Boden mit 4 Schrauben. Das mattschwarze Blatt hat arabische Ziffern von 1 bis 12. Diese sind mit Tritium belegt, gekennzeichnet durch ein kleines T direkt über 6 Uhr, sowie „SWISS" am Zifferblattrand unter der 30. Ein großes 3H-Zeichen befindet sich zentriert im unteren Teil des Zifferblattes. Eine Skala am Blattrand wechselt zwischen Stricheinteilungen für jede Minute und arabischen Ziffern für die 5-Minuten-Markierungen bei 5, 10, 15 bis 60. Über jeder Zahl am Rand des Zifferblattes ist ein kleines Quadrat zu sehen.
Die beiden Hilfszifferblätter sind rund. Die Sekunden-Skala mit einer Stricheinteilung von 1 bis 60 (20, 40, 60 beziffert) befindet sich auf neun Uhr. Der 30-Minuten-Zähler gegenüber zeigt die Zahlen 10, 20 und 30. Alle Skalen sind weiß.
Die weißen Zeiger sind mit Tritium beschichtet. Die Spitze des Chronografen-Zeigers ist dreieckig. Die Zeiger der zwei Hilfszifferblätter sind unterschiedlich geformt.
Anstelle der üblichen 60 hat die schwarze Lünette eine silberne Dreiecksmarkierung, ansonsten Minutenstriche von 1 bis 59. Die 5-Minuten-Markierungen sind über die ganze Breite der Lünette gezogen und beziffert mit 15, 30, 45.
Das Uhrglas aus Kunststoff ist mit einer Dichtung versehen.
Das Gehäuse hat abgerundete Kanten mit einer Bandanstoßbreite von 20 mm. Der Gehäusedeckel zeigt einen kreisförmigen Schliff.
Die gefräste NATO-Nummer 6645-12-146-5081 ist zentral auf der Rückseite positioniert. Eine Gravur ohne Bundeswehr ist selten und bei Uhren der Marine oder von Sondereinheiten zu beobachten.
Das Valjoux signierte Uhrwerk hat eine unsignierte Brücke. Das originale Unterlageband aus Leder war bei dieser frühen Version unsigniert.

Steel screw-back case with 4 screws. The matt black dial has Arabic numerals from 1 to 12, which are coated with tritium, marked by a small T directly above the 6 o'clock position, and "SWISS" on the edge of the dial below the 30-minute marker. A large 3H symbol is centered in the lower part of the dial. A scale on the edge of the dial alternates between graduations for each minute and Arabic numerals for the five-minute markers at 5, 10, 15, to 60. A small square is positioned behind each number on the edge of the dial.
The two subdials have concentric circles. The seconds scale with graduation from 1 to 60 (numbered 20, 40, 60) is located at the nine o'clock position. Opposite is the 30-minute counter, numbered 10, 20, 30. All scales are printed in white.
The white hands are coated with tritium. The chronograph hand has a delta-shaped tip. The hands of the two subdials are differently shaped.
The matt black bezel has a silver triangular marking, otherwise 60 minute markers from one to 59. The five-minute markers are numbered 15, 30, 45 across the entire width of the bezel.
The plastic crystal is fitted with a seal.
The case has rounded edges with a lug width of 20 mm (0.79 in). Concentric circular grinding is visible on the container cover.
The milled NATO number 6645-12-146-5081 is positioned at the center of the back. Engraving without a Bundeswehr mark is rare and can be seen on naval watches or special units.
The Valjoux signed movement has an unsigned bridge. The original leather strap with retracted back cover protection was an early unsigned version of this watch.

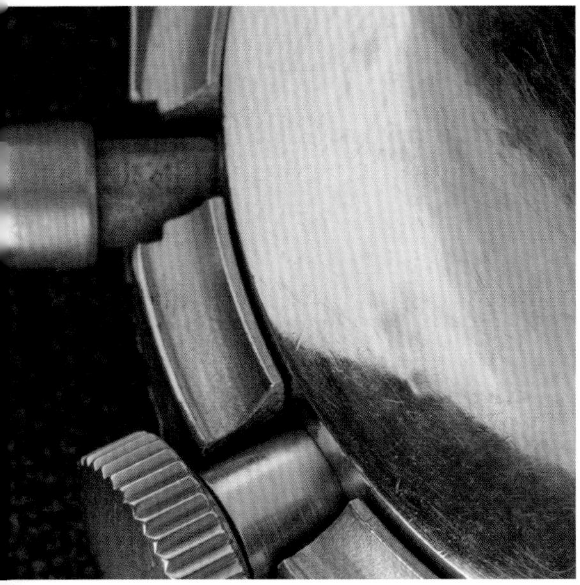

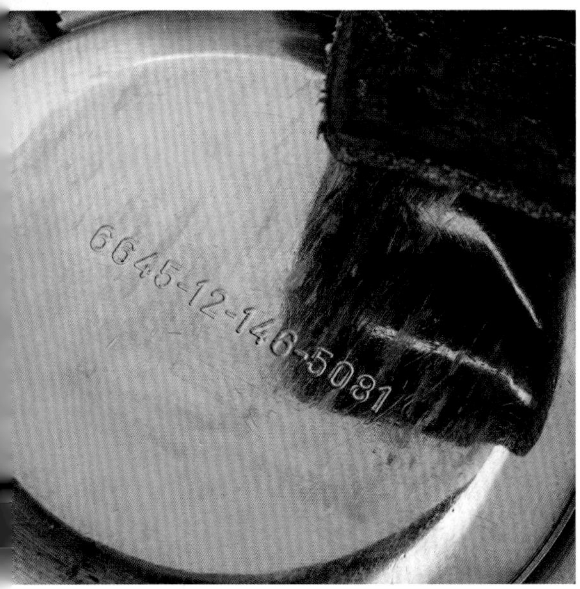

HEUER

HEUER

ref. 1550 SG Medium 3H

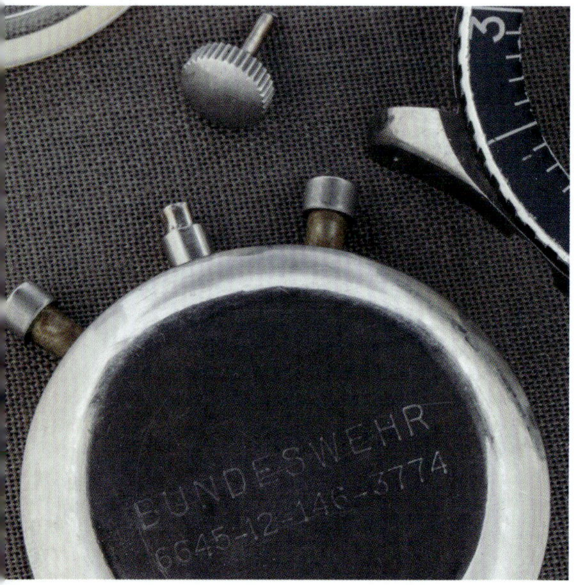

Stahlgehäuse, verschraubter Boden mit 4 Schrauben. Das Zifferblatt ist mattschwarz, mit arabischen Ziffern von 1 bis 12. Diese sind mit Tritium beschichtet, was durch ein kleines T direkt über der 6 gekennzeichnet ist. Am Zifferblattrand unter der 30 steht „SWISS". Ein mittelgroßes 3H-Zeichen befindet sich zentriert im unteren Teil des Zifferblattes. Eine Skala am Zifferblattrand wechselt zwischen Stricheinteilungen für jede Minute und arabischen Ziffern für die 5-Minuten-Markierungen bei 5, 10, 15 bis zur 60. Über jeder Zahl am Rand des Zifferblattes ist ein kleines Quadrat positioniert.
Beide Hilfszifferblätter sind kreisförmig. Auf 9 Uhr befindet sich die Sekunden-Skala mit einer Stricheinteilung von 1 bis 60 (20, 40, 60 beziffert). Der 30-Minuten-Zähler gegenüber zeigt die Zahlen 10, 20 und 30. Alle Skalen sind weiß.
Die weißen Zeiger sind mit Tritium beschichtet. Die Spitze des Chronografen-Zeigers ist dreieckig. Die Zeiger der zwei Hilfszifferblätter sind unterschiedlich geformt.
Anstelle der üblichen 60 hat die schwarze Lünette eine silberne Dreiecksmarkierung, ansonsten Minutenstriche von 1 bis 59. Die 5-Minuten-Markierungen sind über die ganze Breite der Lünette gezogen und beziffert mit 15, 30, 45.
Das Uhrglas aus Kunststoff ist mit einer Dichtung versehen. Das Gehäuse hat abgerundete Kanten mit einer Bandanstoßbreite von 20 mm. Der Rückendeckel zeigt einen kreisförmigen Schliff.
Die gravierte NATO-Nummer 6645-12-146-3774 ist zentral auf der Rückseite positioniert. Darüber befindet sich die Gravur „BUNDESWEHR".
Die Brücke des mit Valjoux signierten Uhrwerks ist unsigniert.
Das originale Unterlageband aus Leder war mit der Versorgungsnummer 6645-12-145-6415 und Bund gekennzeichnet.

Steel screw-back case with 4 screws. The matt black dial has Arabic numerals from 1 to 12, which are covered with tritium, marked by a small T directly above the 6 o'clock position, and "SWISS" on the edge of the dial below the 30-minute marker. A medium-sized 3H symbol is centered in the lower part of the dial. A scale on the edge of the dial alternates between line divisions for each minute and Arabic numerals for the five-minute markers at 5, 10, 15, to 60. A small square is positioned behind each number on the edge of the dial.
The two subdials have concentric circles. The seconds scale with a graduation from 1 to 60 (numbered 20, 40, 60) is located at the nine o'clock position. Opposite is the 30-minute counter, numbered 10, 20, 30. All scales are printed in white.
The white hands are coated with tritium. The chronograph hand has a delta-shaped tip. The hands of the two subdials are differently shaped.
The matt black bezel has a triangular silver marking, otherwise 60 minute markers from one to 59. The five-minute markers are numbered 15, 30, 45 across the entire width of the bezel.
The plastic crystal is fitted with a seal.
The case has rounded edges with a lug width of 20 mm (0.79 in). Concentric circular grinding is visible on the container lid.
The engraved NATO number 6645-12-146-3774 is positioned at the center on the back. The "BUNDESWEHR" engraving is located above it.
The Valjoux signed movement has an unsigned bridge. The original leather strap with retracted back cover protection was labeled with the supply number 6645-12-145-6415 and "BUND."

HEUER

HEUER

ref. 1550 SG Small 3H

Stahlgehäuse, verschraubter Boden mit 4 Schrauben. Das Zifferblatt ist mattschwarz, mit arabischen Ziffern von 1 bis 12. Diese sind mit Tritium beschichtet, was durch ein kleines T direkt über der 6 gekennzeichnet ist. Am Zifferblattrand unter der 30 steht „SWISS". Ein kleines 3H-Zeichen befindet sich zentriert im unteren Teil des Zifferblattes. Eine Skala am Zifferblattrand wechselt zwischen Stricheinteilungen für jede Minute und arabischen Ziffern für die 5-Minuten-Markierungen bei 5, 10, 15 bis zur 60. Über jeder Zahl am Rand des Zifferblattes ist ein kleines Quadrat positioniert.
Beide Hilfszifferblätter sind kreisförmig. Auf 9 Uhr befindet sich die Sekunden-Skala mit einer Stricheinteilung von 1 bis 60 (20, 40, 60 beziffert). Der 30-Minuten-Zähler gegenüber zeigt die Zahlen 10, 20 und 30. Alle Skalen sind weiß.
Die weißen Zeiger sind mit Tritium beschichtet. Die Spitze des Chronografen-Zeigers ist dreieckig. Die Zeiger der zwei Hilfszifferblätter sind unterschiedlich geformt.
Anstelle der üblichen 60 hat die schwarze Lünette eine silberne Dreiecksmarkierung, ansonsten Minutenstriche von 1 bis 59. Die 5-Minuten-Markierungen sind über die ganze Breite der Lünette gezogen und beziffert mit 15, 30, 45.
Das Uhrglas aus Kunststoff ist mit einer Dichtung versehen.
Das Gehäuse hat abgerundete Kanten mit einer Bandanstoßbreite von 20 mm. Der Rückendeckel zeigt einen kreisförmigen Schliff.
Die gravierte NATO-Nummer 6645-12-146-3774 ist zentral auf der Rückseite positioniert. Darüber befindet sich die Gravur „BUNDESWEHR".
Die Brücke des mit Valjoux signierten Uhrwerks ist unsigniert.
Ein mattiertes Stahlband mit offener Gliederstruktur ist durch einen Auszug-Mechanismus in der Schließe verstellbar. Die Anstöße sind genau angepasst und ebenfalls mattiert.

Steel screw-back case with 4 screws. The matt black dial has Arabic numerals from 1 to 12, which are coated with tritium, marked by a small T directly above the 6 o'clock position, and "SWISS" on the edge of the dial below the 30-minute marker. A small 3H symbol is centered in the lower part of the dial. A scale on the edge of the dial alternates between graduations for each minute and Arabic numerals for the five-minute markers at 5, 10, 15, to 60. A small square is positioned behind each number on the edge of the dial.
The two subdials have concentric circles. The seconds scale with graduation from 1 to 60 (numbered 20, 40, 60) is located at the nine o'clock position. Opposite is the 30-minute counter, numbered 10, 20, 30. All scales are printed in white.
The white hands are coated with tritium. The chronograph hand has a delta-shaped tip. The hands of the two subdials are differently shaped.
The matt black bezel has a triangular silver marking, otherwise 60 minute markers from one to 59. The five-minute markers are numbered 15, 30, 45 across the entire width of the bezel.
The plastic crystal is fitted with a seal.
The case has rounded edges with a strap lug width of 20 mm (0.79 in). Concentric circular grinding is visible on the container lid. The engraved NATO number 6645-12-146-3774 is positioned in the center of the back. The "BUNDESWEHR" engraving is located above it.
The Valjoux signed movement has an unsigned bridge.
A matt-finished steel bracelet with an open link structure can be adjusted using a pull-out mechanism in the clasp. The lugs are precisely adjusted and also matt finished.

HEUER

HEUER

HEUER

ref. 1550 SG missing 3H

Stahlgehäuse, verschraubter Boden mit 4 Schrauben. Das Zifferblatt ist mattschwarz, mit arabischen Ziffern von 1 bis 12. Diese sind mit Tritium beschichtet, was durch ein kleines T direkt über der 6 gekennzeichnet ist. Am Zifferblattrand unter der 30 steht „SWISS“. Eine Skala am Zifferblattrand wechselt zwischen Stricheinteilungen für jede Minute und arabischen Ziffern für die 5-Minuten-Markierungen bei 5, 10, 15 bis zur 60. Über jeder Zahl am Rand des Zifferblattes ist ein kleines Quadrat positioniert.
Beide Hilfszifferblätter sind kreisförmig. Auf 9 Uhr befindet sich die Sekunden-Skala mit einer Stricheinteilung von 1 bis 60 (20, 40, 60 beziffert). Der 30-Minuten-Zähler gegenüber zeigt die Zahlen 10, 20 und 30. Alle Skalen sind weiß.
Die weißen Zeiger sind mit Tritium beschichtet. Die Spitze des Chronografen-Zeigers ist dreieckig. Die Zeiger der zwei Hilfszifferblätter sind unterschiedlich geformt.
Anstelle der üblichen 60 hat die schwarze Lünette eine silberne Dreiecksmarkierung, ansonsten Minutenstriche von 1 bis 59. Die 5-Minuten-Markierungen sind über die ganze Breite der Lünette gezogen und beziffert mit 15, 30, 45.
Das Uhrglas aus Kunststoff ist mit einer Dichtung versehen.
Das Gehäuse hat abgerundete Kanten mit einer Bandanstoßbreite von 20 mm. Der Rückendeckel zeigt einen kreisförmigen Schliff. Die gravierte NATO-Nummer 6645-12-146-5081 ist zentral auf der Rückseite positioniert. Eine Gravur ohne Bundeswehr-Zeichen ist selten und bei Uhren der Marine oder von Sondereinheiten zu beobachten.
Die Brücke des mit Valjoux signierten Uhrwerks ist unsigniert.
Das originale Unterlageband aus Leder war mit der Versorgungsnummer 6645-12-145-6415 und Bund gekennzeichnet.
Diese Variante war auch bei den norwegischen Fliegern und anderen Militär-Piloten im Europäischen Raum im Einsatz.

Steel screw-back case with 4 screws. The matt black dial has Arabic numerals from 1 to 12, which are coated with tritium, marked by a small T directly above the 6 o'clock position, and "SWISS" on the edge of the dial below the 30-minute marker.
A scale on the edge of the dial alternates between graduations for each minute and Arabic numerals for the five-minute markers at 5, 10, 15, to 60. A small square is positioned behind each number on the edge of the dial.
The two subdials have concentric circles. The seconds scale with graduation from 1 to 60 (numbered 20, 40, 60) is located at the nine o'clock position. Opposite is the 30-minute counter, numbered 10, 20, 30. All scales are printed in white.
The white hands are coated with tritium. The chronograph hand has a delta-shaped tip. The hands of the two subdials are differently shaped.
The matt black bezel has a triangular silver marking instead of the 60 otherwise minute markers from one to 59. The five-minute markers are numbered 15, 30, and 45 across the entire width of the bezel.
The plastic crystal is fitted with a seal.
The case has rounded edges with a strap lug width of 20 mm (0.79 in). Concentric circular grinding is visible on the container lid.
The milled NATO number 6645-12-146-5081 is positioned at the center on the back. Engraving without a Bundeswehr mark is rare and can be seen on naval watches or special units.
The Valjoux signed movement has an unsigned bridge. The original leather strap with retracted back cover protection was labeled with the supply number 6645-12-145-6415 and "BUND."
This variant was also used by Norwegian airmen and other military pilots.

24 Heures
du Mans
13/14 Juin
1970
CLASSEMENT GÉNÉRAL
1. HERRMANN/ATTWOOD PORSCHE 917
2. LARROUSSE/KAUHSEN PORSCHE 917
3. LINS/DR. MARKO PORSCHE 908
4. POSEY/BUCKNUM FERRARI 512 S
5. DE FIERLAND/WALKER FERRARI 512 S
6. CHASSEUIL/BALLOT-LENA PORSCHE 914-6
7. KREMER/KOOB PORSCHE 911
L'INDICE DE PERFORMANCE
1. LINS/DR. MARKO PORSCHE 908
L'INDICE AU RENDEMENT ÉNERGÉTIQUE
1. LARROUSSE/KAUHSEN PORSCHE 917
COUPE DE GRAND TOURISME
1. CHASSEUIL/BALLOT-LENA PORSCHE 914-6
23
Shell
PORSCHE
PORSCHE
Dr.-Ing. h. c. F. Porsche KG Stuttgart-Zuffenhausen Printed in Germany Juni 1970 Entwurf Atelier Strenger Foto Rusell

MONACO

Der Stadtstaat Monaco liegt an der Côte d'Azur, die sich von Menton bis Saint-Tropez erstreckt. Die Alpen bilden die Kulisse des Küstenstaates und der Strand ist geprägt von malerischen Dörfern wie Cap d'Ail, Beausoleil und Roquebrune-Cap-Martin.
Seit dem ersten Grand Prix am 14. April 1929 haben 82 große Preise von Monaco stattgefunden. Alle zwei Jahre wird ein Oldtimer-Grand-Prix mit ehemals in Monaco gestarteten Autos veranstaltet. Die großen Gladiatoren der Neuzeit sind die Rennfahrer, und es gibt wenige Plätze, wo sich eine Stadt in ihren engen Gassen so aufheizt wie hier in Monaco. Die Strecke fordert alles und die überall präsenten Leitplanken verzeihen keine Fehler. Mit dieser irrealen Geschwindigkeit anbremsen, in die Haarnadelkurve Rascasse hinein, dann vorbei an Start und Ziel den Berg hinauf zum Casino, dann hinunter durch Mirabeau und Loews zum Meer, weiter durch den Tunnel zum Hafen, wo die Autos von der tosenden Menge begrüßt werden, und die Runde sich mit der Swimmingpool-Passage schließt.
Ebenso eindrucksvoll ist auch die Heuer Monaco, der erste rechteckige Chronograph der Welt mit Automatikwerk, der am 3. März 1969 in Genf und New York zeitgleich präsentiert wurde: Bis heute der einzige rechteckige Chronograph mit automatischem Aufzug mittels eines Mikrorotors. Wie kam nun diese „MONACO" von Monte Carlo an den Arm des Hauptdarstellers in dem berühmten Film „Le Mans"? Der 13,8 Kilometer lange Straßenkurs der „24 Heures du Mans" gehört neben dem Formel 1 „Grand Prix von Monaco" und dem „Indianapolis 500"-Rennen in den Vereinigten Staaten zu den berühmtesten Autorennen der Welt. Der authentische Gewährsmann für den Hauptdarsteller Steve McQueen war sein Freund, der Schweizer Rennfahrer Jo Siffert, der von Heuer gesponsert wurde. So kam es, dass Jo Siffert eine wichtige Rolle bei den Dreharbeiten zum Film „Le

The principality of Monaco lies on the Cote d'Azur, which stretches from Menton to St Tropez. The Alps serve as a backdrop to the seafront location; the beach is dotted with picturesque villages including Cap d'Ail, Beausoleil, and Roquebrune-Cap-Martin.
Since the first Grand Prix on April 14, 1929, there have been 82 great races in Monaco. Every two years there is an Oldtimer Grand Prix, where cars from Monaco's glorious past are allowed at the starting line. Race car drivers are the great gladiators of our time, and there are few places where they can do battle that can compare with Monaco; its small streets offer a uniquely exhilarating experience.
The course is demanding and the omnipresent guardrails are unforgiving of the slightest error. Braking with supernatural speed into the hairpin Rascasse curve, racing up the mountain past the starting line to the casino, then down through Mirabeau and Loews to the sea, the cars roar through the tunnel to the port, where they are saluted by cheering crowds as they complete the lap at the Piscine.
Equally impressive is the Heuer Monaco wristwatch, the world's first self-winding rectangular chronograph, which was revealed simultaneously on March 3, 1969, in Geneva and New York simultaneously. To this day, it remains the only square self-winding chronograph with microrotor. How did this Monaco from Monte Carlo land on the arm of the star of the famous film "Le Mans"? The 13.8-kilometre street course of the 24 Heures du Mans is alongside the Formula One Monaco Grand Prix and the Indianapolis 500 in the United States, as one of the most famous car races in the world. The best wingman for the actor Steve McQueen was his friend, Swiss race car driver Jo Siffert, who was sponsored by Heuer. It was this connection that led Jo Siffert to play an important role in the filming of "Le

Rennplakat von Porsche von 1970, welches den eindrucksvollen Erfolg der Marke im Klassement wiederspiegelt.

Porsche race poster showing the impressive success of the brand after the 24h of Le Mans race in 1970.

Das Siegerauto in Le Mans von 1970, der Porsche 917Kdes Teams Porsche Salzburg, gefahren von Hans Herrmann und Richard Atwood, nachts in der Schikane.

The winning car at Le Mans in 1970, the Porsche 917K of the Porsche Salzburg Team, driven by Hans Herrmann and Richard Attwood in the chicane at night.

Mans" spielte. Da er seinem Freund als Vorbild diente, entschied sich McQueen für eine Kopie von Sifferts Rennanzug mit dem prominenten Heuer-Logo auf der Brust. Die gerade auf den Markt gekommene Heuer Monaco Automatik gefiel ihm so gut, dass er beschloss, sie im Film zu tragen. Zu den bereits abgedrehten Rennszenen während des Rennens 1970 drehte man etliche Sequenzen nach und mietete durch die Vermittlung von Jo Siffert die nötigen Rennwagen samt Fahrern wie Derek Bell, Brian Redman, Rolf Stommelen, David Piper und andere. Porsche Salzburg feierte 1970 einen triumphalen Erfolg. Unter der Rennleitung von Dr. Ferdinand Piech und Ernst Strasser konnten sie sich gegen die Konkurrenz von Gulf Porsche, Martini Racing und Ferrari durchsetzen. Sie gewannen mit den Fahrern Richard Attwood und Hans Herrmann in einem Porsche 917K mit der Nummer 23.

Die Modellreihe Monaco umfasst neben der Automatik-Version mit blauem und grauem Ziffernblatt auch Handaufzugsmodelle. Das avantgardistische Design machte die blaue Monaco zu einer absoluten Design-Ikone mit einmaligem Wiedererkennungswert. Gute Exemplare sind rar und die Nachfrage übertrifft das Angebot bei weitem. Dadurch erreicht die Monaco auf den internationalen Märkten Höchstpreise. Seltene Varianten sind die schwarze Monaco mit schwarz beschichtetem Gehäuse sowie frühe Chronomatic Zifferblätter. Weltweit sind in Sammlerkreisen nur wenige Exemplare bekannt.

Die Variante mit Handaufzug und den drei Hilfszifferblättern besticht durch sehr harmonisches Blatt-Design und ist sicher noch sehr unterbewertet. Für TAG Heuer bedeutet die Verbindung zum „Marken-Ambassador" Steve McQueen und dem Film „Le Mans" eine beispiellose, immer noch andauernde Werbung der Monaco-Modellreihe, die immer wieder in der „Classic" Serie von TAG Heuer aufgelegt wurde.

Mans." As he served as a model for his friend, McQueen decided to wear a copy of Siffert's racing uniform with the prominent Heuer logo on the chest. The Heuer Monaco, which had just come to market, impressed McQueen to such an extent that he decided to wear it in the film.

After the race in 1970 a number of scenes were added to the original footage, for which Jo Siffert arranged the necessary race cars and drivers, among whom were Derek Bell, Brian Redman, Rolf Stommelen, David Piper, and others. The Porsche Salzburg Team celebrated a triumphant success in 1970. Under the coaching of Dr. Ferdinand Piech and Ernst Strasser, they were able to trounce their competition from Gulf Porsche, Martini Racing and Ferrari. Their victory was achieved in a Porsche 917K number 23, with Richard Attwood and Hans Herrmann at the wheel.

The Monaco model line included both self-winding and manual-wind versions with blue and gray dials.

The avant-garde design made the blue Monaco a fashion icon. Good specimens have become rare and demand far exceeds supply; this has led to the Monaco reaching a high price on the international market.

Rare versions are the "black Monaco," produced in a very limited run with black-coatged case, and early samples with "CHRONOMATIC" above the Heuer logo on the dial. Only a few such watches are known to exist, all in the hands of collectors.

The manual-wind model with three subdials features a harmonious dial design, and is almost certainly undervalued. For TAG Heuer, Steve McQueen as an ambassador of the brand and the film "Le Mans" serve as a incomparable examples of enduring promotion for the Heuer Monaco series, which experienced a new production run as TAG Heuer's "Classic" Series.

23
Shell
PORSCHE
Shell

MONACO

ref. 1133 B

Gehäuse mit einem gebürsteten quadratischen Bereich um das gewölbte Glas mit leicht gebauchten Seiten. Das metallischblaue Zifferblatt mit glänzender Oberfläche und weißer Schrift hat zwei weiße 30-Minuten- und 12-Stunden-Hilfszifferblätter. Der Modellname „MONACO" steht über dem Heuer-Logo unter 12 Uhr, „AUTOMATIC CHRONOGRAPH" über dem silber gerahmten Datumsfenster bei 6 Uhr.
Polierte Stundenindizes aus Metall mit äußeren Abschrägungen sind an der Außenseite der runden 1/5-Sekunden-Skala mit roten Stundenindizes und kleinen Leuchtpunkten platziert. Ein großer Stahlindex mit roter Bedruckung steht bei 12 Uhr. Eckige gebürstete Zeiger aus Metall haben Leuchtmasse-Einlagen. Der rote Chronographen-Sekundenzeiger reicht bis zur 1/5-Sekunden-Skala. Die Krone ist auf der linken Seite angebracht und mit „HEUER" signiert. Die gerillten Chronographen-Drücker sind auf der rechten Seite platziert und auf der Rückseite voll sichtbar. Der Boden hat einen Bürstenschliff mit polierten abgeschrägten Kanten und zeigt einen Längsschliff mit der „TOOL No. 033", eingraviert über dem Heuer-Logo. Das Uhrwerk ist ein Kaliber 11 und auf der Brücke mit „CAL. 11"signiert.

This iconic automatic Monaco case incorporates a brushed rectangular area with slightly convex sides around the domed crystal. The metallic blue dial with polished surface has white-printed markings and two white 30-minute and 12-hour recording subdials. The model name "MONACO" is positioned over the Heuer logo below 12 o'clock. "AUTOMATIC CHRONOGRAPH" is above the silver-framed date window at 6 o'clock. Polished metal hour markers with external bevels are placed on the outside of the round 1/5-second track, with red markers for the hours and small luminous dots. A large steel marker with red print is at 12 o'clock.
The squared brushed metal hands have luminous inserts. The red chronograph hand reaches the 1/5-second track. The crown is integrated on the left of the case and signed "HEUER." The fluted chronograph pushers are on the right, fully visible from the back. The case back has a brushed finish with polished beveled rim and features longitudinal grinding with "TOOL NO. 033" (Monaco-specific opening tool) engraved over the Heuer logo.
The movement is a Caliber 11, and is signed "CAL.11" on the bridge.

MONACO
HEUER
AUTOMATIC
CHRONOGRAPH
SWISS

MONACO
HEUER
AUTOMATIC
CHRONOGRAPH
SWISS
SPORTS TIMING
HEUER
51
CARELLO
Shell

MONACO

ref. 1133 G

Gehäuse mit verschraubtem Boden und einem gebürsteten rechteckigen Bereich um das gewölbte Glas mit leicht gebauchten Seiten. Das graue Zifferblatt mit weißer Schrift hat zwei schwarze 30-Minuten- und 12-Stunden-Hilfszifferblätter.
Der Modellname „MONACO“ steht über dem Heuer-Logo unter 12 Uhr, „AUTOMATIC CHRONOGRAPH“ über dem silber gerahmten Datumsfenster bei 6 Uhr. Polierte Stundenindizes aus Metall mit äußeren Abschrägungen sind an der Außenseite der runden 1/5-Sekunden-Skala mit roten Stundenindizes und kleinen Leuchtpunkten platziert. Ein großer Stahlindex mit roter Bedruckung steht bei 12 Uhr. Beim 30-Minuten-Hilfszifferblatt sind die Zahlen 5, 30 und 25 auf gleicher Höhe mit den Zahlen 2, 12 und 10 im 12-Stunden-Hilfszifferblatt. Das Gleiche gilt für die Zahlen 10, 15 und 20 sowie 4, 6 und 8.
Gebürstete Zeiger aus Metall mit roten Leuchtmasse-Einlagen sowie roten Spitzen. Der rote Chronographen-Sekundenzeiger reicht bis zur 1/5-Sekunden-Skala.
Die Krone ist auf der linken Seite integriert und mit „HEUER“ signiert. Die gerillten Drücker sind rechts und auf der Rückseite voll sichtbar. Der Boden hat einen Bürstenschliff mit polierten abgeschrägten Kanten und zeigt einen Längsschliff mit der „TOOL No. 033“, eingraviert über dem Heuer-Logo. Das Uhrwerk ist ein Kaliber 11 und mit „CAL. 11“ auf der Brücke signiert.

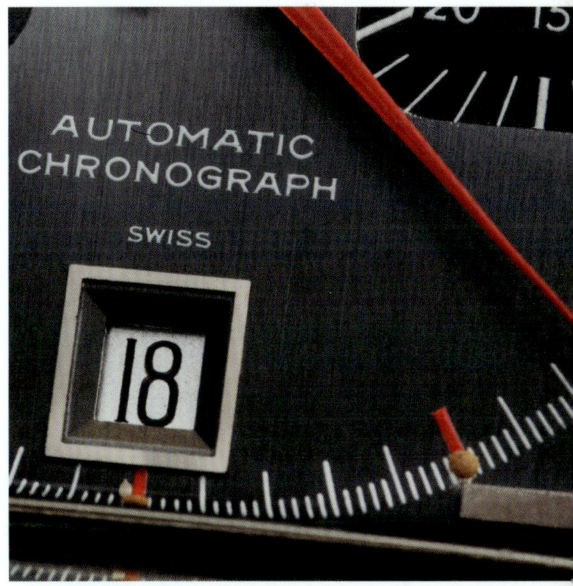

The screw-back case incorporates a brushed rectangular area around the domed crystal with slightly convex sides. The gray dial has white print and two black 30-minute and 12-hour recording subdials.
The model name "MONACO" is positioned over the Heuer logo below 12 o'clock. "AUTOMATIC CHRONOGRAPH" is above the silver-framed date window at 6 o'clock. Polished metal hour indices with external bevels are placed on the outside of the round 1/5-second track with red markers for the hours and small luminous dots.
A large steel marker with red print is at 12 o'clock. On the 30-minute subdial, the numbers 5, 30, and 25 are at the same height as the 2, 12, and 10 on the 12-hour subdial, as are the numbers 10, 15, and 20 and 4, 6, and 8. The brushed metal hands have red and luminous inserts as well as red tips. The red chronograph hand reaches the 1/5-second track.
The crown is integrated on the left of the case and signed "HEUER." The fluted chronograph pushers are on the right, fully visible from the back. The case back has a brushed finish with polished beveled rim and features longitudinal grinding with "TOOL NO. 033" engraved above the Heuer logo.
The movement is a Caliber 11, and is signed "CAL. 11" on the bridge.

McQUEEN
DRIVES
PORSCHE
12 HOURS OF SEBRING 1970
Prototype winner and 2nd overall on Porsche 908
HOLTVILLE RACE 1970
Overall winner on Porsche 908
PHOENIX RACE 1970
Overall winner on Porsche 908
48
Gulf
PORSCHE
Dr.-Ing. h. c. F. Porsche KG Stuttgart-Zuffenhausen Printed in Germany Mai 1970

Gulf
Gulf
2
MONACO
HEUER
AUTOMATIC
CHRONOGRAPH
SWISS
18

MONACO
HEUER
AUTOMATIC
CHRONOGRAPH
SWISS
18
HEUER

MONACO

ref. 1133 B „Steve McQueen"

Gehäuse mit einem gebürsteten rechteckigen Bereich um das gewölbte Glas mit leicht gebauchten Seiten. Das himmelblaue Zifferblatt mit weißer Schrift hat zwei weiße 30-Minuten- und 12-Stunden-Hilfszifferblätter.
Der Modellname „MONACO" steht über dem Heuer-Logo unter 12 Uhr, „AUTOMATIC CHRONOGRAPH" über dem silber gerahmten Datumsfenster bei 6 Uhr. Polierte Stundenindizes aus Metall mit äußeren Abschrägungen sind an der Außenseite der runden 1/5-Sekunden-Skala mit roten Stundenindizes und kleinen Leuchtpunkten platziert. Ein großer Stahlindex mit roter Bedruckung steht bei 12 Uhr. Beim 30-Minuten-Hilfszifferblatt sind die Zahlen 5, 30 und 25 auf gleicher Höhe mit den Zahlen 2, 12 und 10 im 12-Stunden-Hilfszifferblatt. Das Gleiche gilt für die Zahlen 10, 15 und 20 sowie 4, 6 und 8. Gebürstete Zeiger aus Metall mit roten Leuchtmasse-Einlagen sowie roten Spitzen. Der rote Chronographen-Sekundenzeiger reicht bis zur 1/5-Sekunden-Skala.
Die Krone ist auf der linken Seite integriert und mit „HEUER" signiert. Die gerillten Drücker sind rechts und auf der Rückseite voll sichtbar. Der Boden hat einen Bürstenschliff mit polierten, abgeschrägten Kanten und zeigt einen Längsschliff mit der Gravur „TOOL No. 033" über dem Heuer-Logo. Das Uhrwerk ist ein Kaliber 11 und mit „CAL. 11" auf der Brücke signiert.

The case incorporates a brushed rectangular area around the domed crystal with slightly convex sides. The sky-blue dial has white print and two white 30-minute and 12-hour recording subdials. The model name "MONACO" is positioned over the Heuer logo below 12 o'clock. "AUTOMATIC CHRONOGRAPH" is above the silver-framed date window at 6 o'clock. Polished metal hour markers with external bevels are placed on the outside of the round 1/5-second track with red markers for the hours and small luminous dots. A large steel marker with red print is at 12 o'clock. On the 30-minute subdial the numbers 5, 30, and 25 are at the same height as the 2, 12, and 10 on the 12-hour subdial, as are the numbers 10, 15, and 20 and 4, 6, and 8. The brushed metal hands have red and luminous inserts as well as red tips. The red chronograph hand reaches the 1/5-second track. The crown is integrated on the left side of the case and signed "HEUER." The fluted chronograph pushers are on the right, fully visible from the back. The case back has a brushed finish with polished beveled rim and features longitudinal grinding with "TOOL NO. 033" engraved above the Heuer logo.
The movement is a Caliber 11, and is signed "CAL. 11" on the bridge.

MONACO
HEUER
SWISS

MONACO

ref. 73633 G

Dieses Gehäuse mit verschraubtem Boden hat einen gebürsteten quadratischen Rahmen um das gewölbte Glas mit leicht gebauchten Seiten. Das metallischgraue Zifferblatt mit weißer Schrift hat drei graue Hilfszifferblätter, die Sekunden, Minuten und Stunden zählen.
„MONACO“ steht über dem Heuer-Logo unter 12 Uhr, „SWISS“ über 6 Uhr. Polierte Stundenindizes aus Metall mit äußeren Abschrägungen sind an der Außenseite der runden 1/5-Sekunden-Skala mit roten Stundenindizes und kleinen Leuchtpunkten platziert. Ein großer Stahlindex mit roter Bedruckung steht bei 12 Uhr. Beim 30-Minuten-Hilfszifferblatt sind die Zahlen 5, 30 und 25 auf gleicher Höhe mit den Zahlen 10, 60 und 50 im 60-Sekunden-Hilfszifferblatt. Das Gleiche gilt für die Zahlen 10, 15 und 20 sowie 20, 30 und 40.
Gebürstete Zeiger aus Metall mit roten und Leuchtmaterial-Einschlüssen haben rote Spitzen. Der rote Chronographen-Sekundenzeiger reicht bis zur 1/5-Sekunden-Skala.
Die Krone ist mit „HEUER“ signiert, rechts zwischen den gerillten Drückern integriert und auf der Rückseite sichtbar. Der Rückdeckel hat einen Bürstenschliff mit polierten abgeschrägten Kanten und zeigt einen Längsschliff mit der Gravur „TOOL No. 033“ (spezielles Monaco-Öffnungsgerät) über dem Heuer-Logo. Das Uhrwerk ist ein Valjoux Kaliber 7736 und mit „7736 R“ unter der Unruh signiert.

The manual-wind screw-back case incorporates a brushed rectangular area around the domed crystal with slightly convex sides. The metallic gray dial with polished surface has white print and three grey subdials recording seconds, minutes, and hours.
"MONACO" is positioned over the Heuer logo below 12 o'clock and "SWISS" above 6 o'clock. Polished metal hour markers with external bevels are placed on the outside of the round 1/5-second track, with red markers for the hours and small luminous dots. A large steel marker with red print is at 12 o'clock. On the 30-minute subdial the numbers 5, 30, and 25 are at the same height as the 10, 60, and 50 on the second subdial, as are the numbers 10, 15, and 20 and 20, 30, and 40.
The brushed metal hands have red and luminous inserts as well as red tips. The red chronograph hand reaches the 1/5-second track at the edge of the dial.
The crown is signed "HEUER" and on the right between the fluted pushers, which are visible from the back. The case back has a brushed finish with polished beveled rim and features longitudinal grinding with "TOOL No. 033" (Monaco-specific opening tool) engraved over the Heuer logo.
The movement is a Valjoux 7736, and is signed "7736 R" under the balance wheel.

MONACO

ref. 73633 B

Das Gehäuse hat einen gebürsteten quadratischen Bereich um das gewölbte Glas mit leicht gebogenen Seiten. Das lavendelblaue Zifferblatt mit weißer Schrift hat drei weiße Hilfszifferblätter, die Sekunden, 30 Minuten und 12 Stunden zählen. Der Modellname „MONACO" steht über dem Heuer-Logo unter 12 Uhr, „SWISS" über 6 Uhr.
Polierte Stundenindizes aus Metall mit äußeren Abschrägungen sind an der Außenseite der runden 1/5-Sekunden-Skala mit roten Stundenindizes und kleinen Leuchtpunkten platziert. Ein großer Stahlindex mit roter Bedruckung steht bei 12 Uhr. Beim 30-Minuten-Hilfszifferblatt sind die Zahlen 5, 30 und 25 auf gleicher Höhe mit den Zahlen 10, 60 und 50 im 60-Sekunden-Hilfszifferblatt. Das Gleiche gilt für die Zahlen 10, 15 und 20 sowie 20, 30 und 40. Die gebürsteten Zeiger aus Metall mit roten und Leuchtmasse-Einlagen haben rote Spitzen. Der rote Chronographen-Sekundenzeiger reicht bis zur 1/5-Sekunden-Skala. Die Krone ist mit „HEUER" signiert, rechts zwischen den gerillten Drückern integriert, auf der Rückseite sichtbar. Die Bodenrückseite bei dieser ungetragenen Uhr trägt noch das original Heuer-Etikett.
Das Uhrwerk ist ein Valjoux Kaliber 7736 und mit „7736 R" unter der Unruh signiert.

The manual-wind case incorporates a brushed rectangular area around the domed crystal with slightly convex sides. The lavender-blue dial has white print and three white subdials recording seconds, 30 minutes, and 12 hours. The model name "MONACO" is positioned over the Heuer logo below 12 o'clock, and "SWISS" above 6 o'clock.
Polished metal hour markers with external bevels are placed on the outside of the round 1/5-second track, with red indices for the hours and small luminous dots. A large steel marker with red print is at 12 o'clock. On the 30-minute subdial the numbers 5, 30, and 25 are at the same height as the 10, 60, and 50 on the second subdial, as are the numbers 10, 15, and 20 and 20, 30, and 40.
The brushed metal hands have red and luminous inserts as well as red tips. The red chronograph hand reaches the 1/5-second track at the edge of the dial. The crown is signed "HEUER" and integrated on the right between the fluted pushers, which are visible from the back. The case back has an original Heuer sticker on this unworn watch.
The movement is a Valjoux 7736, and is signed "7736 R" under the balance wheel.

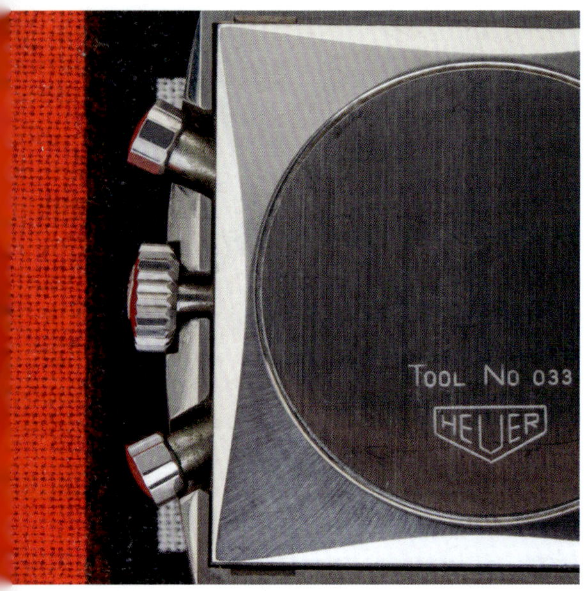

MARTINI
38
MONACO
HEUER
SWISS

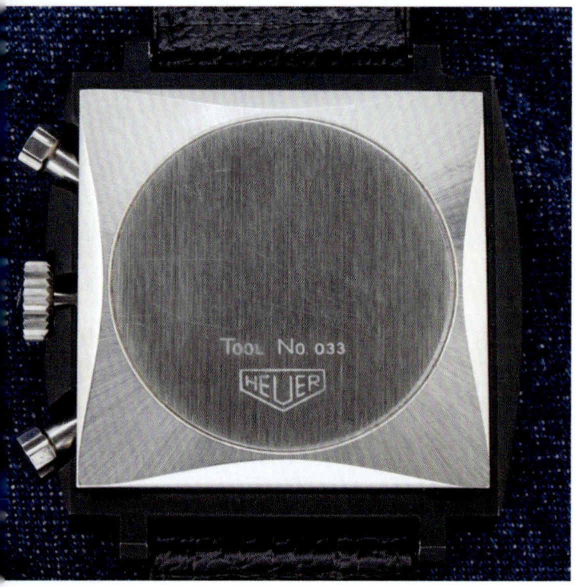

MONACO

ref. 74031 NC

Schwarz beschichtetes Gehäuse mit Handaufzug und einem quadratischen Bereich um das gewölbte Glas mit leicht ovalen Seiten. Das schwarze Zifferblatt mit matter Oberfläche und weißer Schrift hat zwei schwarze 30-Minuten- und 12-Stunden-Hilfszifferblätter. Das einzigartige quadratische Zifferblatt ist perfekt in die Form des Gehäuses eingebettet. Der Modellname „MONACO" steht über dem Heuer-Logo unter 12 Uhr, „SWISS" über dem weißgerahmten Datumsfenster bei 6 Uhr.
Lange weiße Stundenindizes sind auf der Innenseite der runden 1/5-Sekunden-Skala platziert. Ein Doppelindex steht bei 12 Uhr. Weiß bemalte Zeiger haben große Leuchtmasse-Einlagen. Der orangefarbene Chronographen-Sekundenzeiger reicht bis zur 1/5-Sekunden-Skala. Die Krone ist mit „HEUER" signiert, auf der rechten Seite zwischen den gerillten Chronographen-Drückern integriert und auf der Rückseite voll sichtbar. Der Boden hat einen Bürstenschliff mit polierten abgeschrägten Kanten und zeigt einen Längsschliff mit der Gravur „TOOL No. 033" (spezielles Monaco-Öffnungsgerät) über dem Heuer-Logo. Das Uhrwerk ist ein Valjoux Kaliber 7740 und mit „7740 R" unter der Unruh signiert.

The iconic manual-wind black-coated case incorporates a rectangular area around the domed crystal with slightly convex side faces. The black dial with matt surface has white print and black 30-minute and 12-hour subdials. The unique square dial is perfectly embedded in the case. The model name "MONACO" is positioned over the Heuer logo below 12 o'clock, and "SWISS" is above the white-framed date window at 6 o'clock.
Long white luminous hour indices are placed on the inside of the rounded 1/5-second track, with a twin index at 12 o'clock. The white-painted hands have large luminous inserts. The orange chronograph hand reaches the 1/5-second track at the dial.
The crown is signed "HEUER" and integrated on the right between the fluted chronograph pushers, which are fully visible from the back. The case back has a brushed finish with polished beveled rim and shows longitudinal grinding with "TOOL No. 033" (Monaco-specific opening tool) engraved over the Heuer logo.
The movement is a Valjoux 7740, and is signed "7740 R" under the balance wheel.

MONACO
HEUER
SWISS
18

Heuer stoppt Ferrari.

Heuer produziert seit mehr als 110 Jahren Stoppuhren. Inzwischen ist Heuer der größte Stoppuhrenhersteller der Welt geworden.

Berühmte Rennteams wie Ferrari, Alfa Romeo, BRM und Surtees verwenden ausschließlich Heuer-Stoppuhren für die Zeitmessung.

Heuer ist auf allen großen Rennstrecken der Welt zu Hause. Natürlich auch beim „Großen Preis von Österreich". Und wenn Sie sich dort in den Boxen ein bißchen umsehen, werden Sie diese vier Heuer-Modelle öfters sehen:

Chronosplit. Die Uhr, die Niki Lauda trägt. Der erste elektronische Chronograph auf der Welt mit doppelter Digitalanzeige. Flüssigkristallanzeige (LCD) für die Tageszeit und Datum. Leuchtdioden (LED) für die Kurzzeitmessung.

Montreal. Die Uhr, die Clay Regazzoni trägt. Chronograph. Automatic. 12-Stunden-Zähler. Datum.

Microsplit. Die kleinste elektronische Stoppuhr der Welt. Mit Digitalablesung auf eine 100stel Sekunde genau. Drei Zeitmeßmethoden. Meßbereich bis 59 Minuten und 59,99 Sekunden.

Trackmaster. Additions-Stoppuhr. 1/10- oder 1/5-Sekunden-Anzeige. Stoßsicherung. Unzerbrechliche Aufzugsfeder. Antimagnetische Spiralfeder. Bruchsicheres Glas.

HEUER

Bei jedem sportlichen Uhren-Fachhändler.

Bezugsquellennachweis
Chronoswiss
A-1014 Wien, Tuchlauben 8, Telefon 63 07 16

MONTREAL

Dieses neue Modell passte sehr gut zu Jack Heuers Wachstumsstrategie. Der amerikanische Markt war seit jeher ein Steckenpferd von ihm und eine Expansion in den Norden der USA lag auf der Hand. In Montreal spricht man zu 80 Prozent französisch und die Stadt gilt als zweitgrößte französischsprachige der Welt. Bereits 1967 fanden hier die Weltausstellung statt und 1976 die Olympischen Sommerspiele, wo Heuer als Zeitmesser aktiv war. Auch die Formel 1 gastiert seit 1978 jedes Jahr in Montreal. Der Große Preis von Kanada findet auf dem Stadtkurs Circuit Gilles Villeneuve auf der Île Notre-Dame, einer Insel im Sankt-Lorenz-Strom, statt. Die neue Modellreihe besticht durch ein neues, modernes, fast kissenförmiges Design. Das Gehäuse mit dem vom Glasrand abfallenden Strahlenschliff und den glänzend polierten Seiten mit konvexem Schwung ist ein Blickfang. Angetrieben wird die Montreal durch ein Kaliber-12-Automatikwerk. Die Chronographen-Drücker befinden sich an der rechten Seite, die Aufzugskrone sitzt links auf Höhe der 9, eingebettet in eine Kerbung, um sie vor Beschädigungen zu schützen.

Alle Modellvarianten besitzen die gleiche Gehäuseform. Neben den Stahlversionen gab es auch welche mit golden oder schwarz eloxierter (PVD Coating), rauer Oberfläche. Bei den Stahlgehäusen konnte man zwischen blauen, schwarzen und weißen Zifferblättern mit jeweils schwarz oder weiß kontrastierenden Hilfszifferblättern wählen.

Das Zifferblatt wurde durch einen schräg stehenden Außenring visuell enorm vergrößert, geschickt war dort eine Pulsations-Skala zwischen 12 und 3 und daran anschließend eine Tachymeter-Skala angebracht worden. Die Montreal-Zifferblätter sind besonders kontrastreich und glänzen daher speziell bei Uhren in ungetragenem Zustand wunderschön. Die Modellreihe blieb bis Ende der 1970er Jahre im Programm und hat eindrucksvoll ihre Rolle als Designvorreiter untermauert.

This new model fitted Jack Heuer's strategy of expansion exceptionally well. He had long been fascinated by the American market and it was only natural that he to expand north beyond the borders of the USA. Eighty percent of the inhabitants of Montreal speak French, and it is the second largest French-speaking city in the world. It hosted the 1967 World's Fair and, even more importantly for Heuer, the 1976 Olympic Games, for which the company provided the timekeeping expertise. Formula One racing has also hosted a race in Montreal every year since 1978. The Canadian Grand Prix takes place on the Circuit Gilles Villeneuve on Île Notre-Dame, an island in the Saint Lawrence River.

The new model line won over consumers with a totally new, modern design. The almost rectangular case is eye-catching, with its sloping polished grind descending from the bezel to the polished convex sides. It is powered by the Calibre 12 automatic movement. The chronograph's pushers are on the right. The winding crown is positioned at 9 o'clock and is embedded in a protective recessed notch.

All variations of the model line have the same case. As well as stell and gold-plated versions, there was also a novel black-anodized (PVD coating) case with rough surface. Customers could choose blue, black, or white dials with either black or white contrasting subdails in the standard steel case.

The dial was significantly optically enlarged by an angled outer ring with skillfully incorporated pulsationer and tachymeter scale. The dials of the Montreal model are high-contrast and sparkle when found in mint condition. This model series remained in the portfolio until the end of the 1970s and impressively confirmed its role as a design innovation.

MONTREAL

ref. 110.503 B

Rechteckig geformtes Gehäuse mit verschraubtem Boden, Strichschliff und poliertem Ring um das Glas. Das Zifferblatt ist indigoblau mit zwei weißen Hilfszifferblättern, gelben 5-Minuten-Markierungen auf dem 30-Minuten- und nummerierten Stundenindizes von 1 bis 12 auf dem Stunden-Hilfszifferblatt. Die äußere 1/5-Sekunden-Skala ist weiß, mit roten Markierungen bei jeder Stunde und auf gleichem Niveau wie die Hilfszifferblätter. Das Zifferblatt ist umgeben von einer blauen Tachymeter- und Pulsations-Skala.
Der Modellname „MONTREAL" steht unter dem Heuer-Logo bei 12 Uhr, „AUTOMATIC CHRONOGRAPH" über dem Datumsfenster bei 6 Uhr. Stundenindizes aus Stahl schließen Leuchtmasse ein und sind am äußeren Ende des Zifferblattes platziert. Ein größerer Index ist bei 12 Uhr, keine Indizes bei 3, 6 und 9 Uhr.
Die roten Zeiger haben große Leuchtmasse-Einlagen. Der rote Chronographen-Sekundenzeiger erreicht die weiße 1/5-Sekunden-Skala am äußeren Rand des Zifferblattes.
Die Krone ist mit „HEUER" signiert. Die Drücker sind rechts und auf der Rückseite sichtbar. Der Boden hat einen Längsschliff und polierte Seiten.
Das Uhrwerk ist ein Kaliber 12 und mit „CAL. 12" auf der Brücke signiert.

Rectangular screw-back case with brushed finish and polished circle around the crystal. The indigo dial has two white subdials with yellow 5-minute marks on the 30-minute and numbered hour markers from 1 to 12 on the hour subdial. The outer 1/5-second track is white with red markings at each hour and at the same level as the subdials. The dial is surrounded by a blue twin tachymeter and pulsometer scale.
The model name "MONTREAL" is positioned below the Heuer logo under 12 o'clock. "AUTOMATIC CHRONOGRAPH" is printed above the date window at 6 o'clock. Steel hour markers including luminous inserts are placed at the outer edge of the dial. A larger marker is at 12 o'clock and markers have been omitted at 3, 6, and 9 o'clock.
The red-painted hands have large luminous inserts. The red chronograph hand reaches the white 1/5-second track on the outer edge of the dial.
The crown is signed "HEUER." The pushers are on the right and visible from the back. The case back has longitudinal grinding with polished sides.
The movement is a Calibre 12, and is signed "CAL.12" on the bridge.

TACHYMETRE
15 PULSATIONS
HEUER
MONTREAL
AUTOMATIC
CHRONOGRAPH
SWISS

HEUER
REF.
unbreakable mainspring
shock resistant
antimagnetic
lever 17 jewels
HEUER
HEUER
CAL.

MONTREAL

ref. 110.503 N

Rechteckig geformtes Gehäuse mit verschraubtem Boden, Strichschliff und poliertem Ring um das Glas. Das Zifferblatt ist schwarz mit zwei weißen Hilfszifferblättern, gelben 5-Minuten-Markierungen am 30-Minuten- und nummerierten Stundenindizes von 1 bis 12 am Stunden-Hilfszifferblatt. Die äußere 1/5-Sekunden-Skala ist weiß mit roten Markierungen bei jeder Stunde und auf gleichem Niveau wie die Hilfszifferblätter. Das Zifferblatt ist umgeben von einer schwarzen Tachymeter- und Pulsations-Skala.
Der Modellname „MONTREAL“ steht unter dem Heuer-Logo bei 12 Uhr, „AUTOMATIC CHRONOGRAPH“ über dem Datumsfenster bei 6 Uhr. Stundenindizes aus Stahl schließen Leuchtmasse ein und sind am äußeren Ende des schwarzen Ziffernblattes platziert. Ein größerer Index ist bei 12 Uhr, keine Indizes bei 3, 6 und 9 Uhr. Die rot bemalten Zeiger haben große Leuchtmasse-Einlagen. Der rote Chronographen-Sekundenzeiger erreicht die weiße 1/5-Sekunden-Skala am äußeren Rand des Zifferblattes. Die Krone ist mit „HEUER“ signiert. Die Drücker sind rechts. Der Boden hat einen Längsschliff und polierte Seiten. Das Uhrwerk ist ein Kaliber 12 und mit „CAL. 12“ auf der Brücke signiert.

Rectangular screw-back case with brushed finish and a polished circle around the crystal. The black dial has two white subdials with yellow 5-minute marks on the 30-minute and numbered hour markers from 1 to 12 on the hour subdial. The outer 1/5-second track is white with red markings at each hour and at the same level as the subdials. The dial is surrounded by a black twin functional tachymeter und pulsometer scale.
The model name "MONTREAL" is positioned below the Heuer logo under 12 o'clock. "AUTOMATIC CHRONOGRAPH" is printed above the date window at 6 o'clock. Steel hour markers including luminous inserts are placed at the outer edge of the black dial. A larger marker is at 12 o'clock and indices have been omitted at 3, 6, and 9 o'clock. The red-painted hands have large luminous inserts. The red chronograph hand reaches the white 1/5-second track on the outer edge of the dial. The crown is signed "HEUER." The pushers are integrated on the right. The case back has longitudinal grinding with polished sides.
The movement is a Calibre 12, and is signed "CAL. 12" on the bridge.

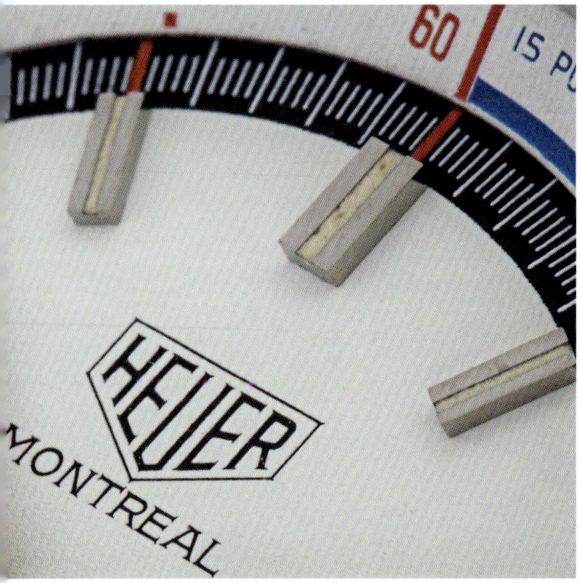

MONTREAL

ref. 110.503 W

Rechteckig geformtes Gehäuse mit verschraubtem Boden, Bürstenschliff und poliertem Ring um das Glas. Das Zifferblatt ist eierschalenfarben, mit zwei schwarzen Hilfszifferblättern, gelben 5-Minuten-Markierungen auf dem 30-Minuten- und nummerierten Stundenindizes von 1 bis 12 auf dem Stunden-Hilfszifferblatt. Die äußere 1/5-Sekunden-Skala ist schwarz mit roten Markierungen bei jeder Stunde und auf dem gleichen Niveau wie die Hilfszifferblätter. Das Zifferblatt ist umgeben von einer eierschalenfarbenen Tachymeter- und Pulsations-Skala.
Der Modellname „MONTREAL" steht unter dem Heuer-Logo bei 12 Uhr, „AUTOMATIC CHRONOGRAPH" über dem Datumsfenster bei 6 Uhr. Stundenindizes aus Stahl schließen Leuchtmasse ein und sind am äußeren Ende des Zifferblattes platziert. Ein größerer Index ist bei 12 Uhr, keine Indizes bei 3, 6 und 9 Uhr. Die rot bemalten Zeiger haben Leuchtmasse-Einlagen. Der rote Chronographen-Sekundenzeiger erreicht die schwarze 1/5-Sekunden-Skala am äußeren Rand des Zifferblattes. Die Krone ist mit „HEUER" signiert. Die Drücker sind rechts und auf der Rückseite sichtbar. Der Boden hat einen Längsschliff und polierte Seiten.
Das Uhrwerk ist ein Kaliber 12 und mit „CAL. 12" auf der Brücke signiert.

Rectangular shaped screw-back case with brushed finish and a polished circle around the crystal. The off-white dial has two black subdials with yellow 5-minute marks on the 30-minute and numbered hour markers from 1 to 12 on the hour subdial. The outer 1/5-second track is black with red markings at each hour and at the same level as the subdials. The dial is surrounded by an off-white twin functional tachymeter and pulsometer scale.
The model name "MONTREAL" is positioned below the Heuer logo under 12 o'clock and "AUTOMATIC CHRONOGRAPH" is printed above the date window at 6 o'clock. Steel hour markers including luminous inserts are precisely placed at the outer end of the off-white dial. A larger marker is at 12 o'clock and markers have been omitted at 3, 6, and 9 o'clock. The red-painted hands have large luminous inserts. The red chronograph hand reaches the black 1/5-second track on the outer edge of the dial. The crown is signed "HEUER." The pushers are on the right and visible from the back. The case back has longitudinal grinding with polished sides. The movement is a Calibre 12, and is signed "CAL. 12" on the bridge.

HEUER
MONTREAL
AUTOMATIC CHRONOGRAPH
SWISS
15 PULSATIONS

MONTREAL

ref. 110.503 BC

Rechteckig geformtes Gehäuse mit verschraubtem Boden, Strichschliff und poliertem Ring um das Glas. Das Zifferblatt ist blau mit weißen Hilfszifferblättern und roten 5-Minuten-Markierungen auf dem 30-Minuten- und nummerierten Stundenindizes von 1 bis 12 am Stunden-Hilfszifferblatt. Die äußere 1/5-Sekunden-Skala ist blau mit weißen Markierungen bei jeder Stunde und auf gleichem Niveau wie die Hilfszifferblätter. Das Zifferblatt ist umgeben von einer blauen Tachymeter- und Pulsations-Skala.
Der Modellname „MONTREAL" steht unter dem Heuer-Logo bei 12 Uhr, „AUTOMATIC CHRONOGRAPH" über dem Datumsfenster bei 6 Uhr. Stundenindizes haben mit Leuchtmasse gefüllte Flächen und sind am äußeren Ende des blauen Ziffernblattes platziert. Die Indizes bei 3, 6 und 9 Uhr sind verkürzt. Die rot bemalten Zeiger haben große Leuchtmasse-Einlagen. Der rote Chronographen-Sekundenzeiger erreicht die blaue 1/5-Sekunden-Skala am äußeren Rand des Zifferblattes. Die Krone mit „HEUER" signiert. Die Drücker sind rechts platziert und auf der Rückseite sichtbar. Der Boden hat einen Längsschliff und polierte Seiten.
Das Uhrwerk ist ein Kaliber 12 und mit „CAL. 12" auf der Brücke signiert.

Rectangular screw-back case with brushed finish and a polished circle around the crystal. The blue dial has white markers with red 5-minute marks on the 30-minute and numbered hour indices from 1 to 12 on the hour subdial. The outer 1/5-second track is blue with white markings at each hour and at the same level as the subdials. The dial is surrounded by a blue twin functional tachymeter and pulsometer scale.
The model name "MONTREAL" is positioned below the Heuer logo under 12 o'clock. "AUTOMATIC CHRONOGRAPH" is printed above the date window at 6 o'clock. Hour markers with luminous inserts are placed at the outer end of the blue dial. Markers have been shortened at 3, 6, and 9 o'clock. The red-painted hands have large luminous inserts. The red chronograph hand reaches the blue 1/5-second track on the outer edge of the dial. The crown is signed "HEUER." The pushers are on the right and visible from the back. The case back has longitudinal grinding with polished sides.
The movement is a Calibre 12, and is signed "CAL.12" on the bridge.

HEUER
MONTREAL
AUTOMATIC
CHRONOGRAPH
SWISS
TACHYMETRE
15 PULSATIONS
gazzoni
Stunden-
HEUER

MONTREAL

ref. 110.501 N

Rechteckig geformtes Gehäuse mit verschraubtem Boden, dickem PVD-Überzug und rauer Oberfläche. Das Zifferblatt ist schwarz, mit zwei schwarzen Hilfszifferblättern. Die äußere 1/5-Sekunden-Skala ist schwarz mit weißen Markierungen bei jeder Stunde und auf dem gleichen Niveau wie die Hilfszifferblätter. Das Zifferblatt ist umgeben von einer schwarzen Tachymeter- und Pulsations-Skala.
Der Modellname „MONTREAL" steht unter dem Heuer-Logo bei 12 Uhr, „AUTOMATIC CHRONOGRAPH" über dem Datumsfenster bei 6 Uhr. Stundenindizes haben mit Leuchtmasse gefüllte Flächen und sind am äußeren Ende des schwarzen Ziffernblattes platziert. Die Indizes bei 3, 6 und 9 Uhr sind verkürzt. Die weißen Zeiger haben große Leuchtmasse-Einlagen. Der weiße Chronographen-Sekundenzeiger erreicht die schwarze 1/5-Sekunden-Skala am äußeren Rand des Zifferblattes. Die Krone ist mit „HEUER" signiert. Die Drücker sind rechts und auf der Rückseite sichtbar. Der Boden hat einen Längsschliff und polierte Seiten.
Das Uhrwerk ist ein Kaliber 12 und mit „CAL. 12" auf der Brücke signiert.

Rectangular screw-back case has a thick PVD coating and rough surface finish.
The dial is black with two black subdials. The outer 1/5-second track is black with white markings at each hour and at the same level as the subdials. The dial is surrounded by a black twin functional tachymeter and pulsometer scale.
The model name "MONTREAL" is positioned below the Heuer logo under 12 o'clock. "AUTOMATIC CHRONOGRAPH" is printed above the date window at 6 o'clock. Hour markers with luminous inserts are placed at the outer end of the black dial. Markers have been shortened at 3, 6, and 9 o'clock. White hands have large luminous inserts. White chronograph hand reaches the black 1/5-second track on the outer edge of the dial. The crown is signed "HEUER." The pushers are integrated on the right and visible from the back. The case back has longitudinal grinding with polished sides.
The movement is a Calibre 12, and is signed "CAL. 12" on the bridge.

TACHYMETRE
15 PULSATIONS
HEUER
MONTREAL
AUTOMATIC
CHRONOGRAPH
SWISS

MONZA
HEUER
AUTOMATIC
CHRONOGRAPH

MONZA

Am 7. September 1975 ging in Monza ein Raunen durch die 200.000 Zuschauer beim Großen Preis von Italien. Der Tag war bewölkt, aber warm. Der Ferrari-Fahrer Clay Regazzoni schoss nach 52 Runden als erster über die Ziellinie. Doch der Drittplatzierte Niki Lauda schrieb an diesem Tag Motorsport-Geschichte, denn er war der neue Formel 1-Weltmeister für Ferrari. Elf lange Jahre waren vergangen, seit John Surtees 1964 den letzten Titel für die Scuderia Ferrari gewonnen hatte.

Luca de Montezemolo als Teamchef, Mauro Forghieri als technischer Direktor, sowie die Fahrer Niki Lauda und Clay Regazzoni hatten 1975 die Geschicke von Ferrari in der Hand. Sie verbesserten den Ferrari 312 T zu einem unschlagbaren Siegerauto. Für den Hauptsponsor Heuer war das ein wichtiger Erfolg, prangte doch das Firmenlogo in großen Buchstaben auf der Karosserie des Ferraris. Eine Bestätigung für Jack Heuers Strategie, den Motorsport zu sponsern: Er hatte auf das richtige Pferd gesetzt. Man war sogar so erfreut im Hause Heuer, dass man diesem Ereignis eine eigene neue Modellreihe widmete. Es sollte eine besondere Kleinserie werden: Sie erschien zur Feier des in Monza von Niki Lauda für Ferrari eroberten Weltmeistertitels 1975, aber auch der Zusammenarbeit zwischen Heuer und Ferrari. Geliefert wurde sie in einer roten, von der Form eines Helms inspirierten Verpackung und in begrenzter Stückzahl.

Gehäusetechnisch ähnlich der Carrera wurde für das Heuer Monza-Modell ein Kaliber 15 mit der kleinen Sekunde auf 10 Uhr verbaut. Der Name „MONZA“ war über dem klassischen Heuer-Logo in der oberen Hälfte des Zifferblattes positioniert, im Gegensatz zu den Varianten der Autavia, Carrera oder Calculator mit Kaliber 15, wo sich der Name jeweils bei 9 Uhr befand. Das klare, gut lesbare Zifferblatt wird von einer Pulso- und Tachymeter-Skala eingerahmt – beides sehr wichtige Bestandteile in der Welt des Motorsports.

On September 7, 1975, a murmur rippled through the 200,000 spectators at the Italian Grand Prix in Monza. The sky was overcast, but the temperature was warm. Ferrari driver Clay Regazzoni crossed the finish line first after 52 laps, but it was Niki Lauda, who finished third, who made motorsport history that day by securing the Formula 1 World Championship for Ferrari.

It had been 11 years since John Surtees had last claimed a title for Scuderia Ferrari in 1964.

Under the leadership of team principal Luca di Montezemolo, technical director Mauro Forghieri, and drivers Niki Lauda and Clay Regazzoni, Ferrari transformed the Ferrari 312 T into an unstoppable force. This championship victory was also significant for Heuer, the team's sponsor, whose logo was prominently displayed on the Ferrari's bodywork.

Heuer's sponsorship strategy had clearly paid off. So pleased was Jack Heuer with the success that, he commemorated the 1975 World Championship with a special limited-edition watch series. This model was designed not only to celebrate Lauda's victory at Monza, but also to mark the collaboration between Heuer and Ferrari. The watch came in red Heuer helmet-inspired packaging and was a tribute to the triumph of Niki Lauda.

The design of the Monza chronograph was based on the Carrera model but featured a chrome or black-coated case. Heuer exclusively used Calibre 15 for the first Monza series, notable for its small second hand at 10 o'clock. Unlike the Autavia, Carrera, or Calculator models, where the Heuer logo appeared at 9 o'clock, the Monza logo was placed above the Heuer logo at the top of the dial. The luminous black dial featured a 1/5 second division, with a red hand for the small second, a white angular 30-minute hand, and black hour and minute hands that tapered to white tips. The red chronograph second hand completed the design.

Jack Heuer mit Ferraris Rennsportleiter Luca de Montezemolo beim Grand Prix 1975.

Jack Heuer with the head of Ferrari's racing division Luca de Montezemolo at the Grand Prix in 1975.

Die Pulsometer-Skala gibt die Anzahl der Pulsschläge pro Minute auf der Basis von 15 an. Die Tachymeter-Skala dient zur Ermittlung der mittleren Geschwindigkeit pro Stunde auf 1 Kilometer oder 1 Meile gerechnet, die dann am Rand des Zifferblatts abgelesen werden kann.
Bei den Monza-Modellen gravierte man auf die Boden-Rückseite den Namen Monza über dem Heuer-Logo, darunter stand die Referenz- und Seriennummer, die ansonsten bei den anderen Modellen meistens zwischen den Bandanstößen bei 6 und 12 zu finden sind. Auf dem Stahlboden konnten die Nummern gut sichtbar angebracht werden, denn das mit Chrom oder schwarz beschichtete Gehäuse war dafür nicht geeignet. Die Krone war unsigniert.
1976 produzierte Heuer eine faltbare Werbebroschüre für die Monza mit Referenz 150.501 (schwarzes PVD-Gehäuse) sowie 150.511 (Chrom). Ein Foto von Niki Lauda mit der Uhr am Handgelenk, sowie eine von ihm signierte Produktempfehlung, ließ keinen Zweifel aufkommen, dass die Monza die Uhr des Weltmeisters war.
Explizit wurde auf die Zusammenarbeit mit Ferrari und die Erfolge des Hauses Heuer als Zeitnehmer im Motorsport hingewiesen.
Die Modelllinie Monza beschrieb man als bahnbrechend in Qualität, Design und Preis und zielte mit dem damals 26-jährigen Lauda auf eine junge sportliche Zielgruppe ab. Außerdem diversifizierte Heuer die neue Produktlinie in:

1. Monza-Sport: schwarz beschichtet mit Lochlederband
2. Lady-Monza: schwarz beschichtet mit „elegantem schwarzweißem Lederband für die Dame“
3. Monza-City: „polierte Metallausführung mit rostfreiem Metallarmband“.

Die Garantiezeit betrug damals schon ein Jahr und der Kaufpreis konnte in Monatsraten abbezahlt werden.
Im darauffolgenden Jahr erweiterte Heuer das Monza-Portfolio um eine Kaliber 12-Variante mit der Referenz 110.511.
Für die Monza-Reihe wurden verschiedene Gehäuse-Versionen verwendet, die sich in Form, Beschichtung und Größe unterschieden. Bei den Gehäusen mit Chrom und schwarzer PVD-Beschichtung fungierte Messing als Basis. Die schwarzen Gehäuse gab es als Varianten mit oder ohne Strichschliff, das Chrom-Pendant nur mit.
Die Rückdeckel variierten bei Gravur und Typographie an Außen-und Innenseite.
Laut meiner Recherche verschwand die Monza-Reihe bereits nach 1979 wieder aus Heuers Produkt-Katalogen, um 1981 als „Economy“-Version ohne „Monza“ am Zifferblatt mit Kaliber 12 wiederzukehren. Die Referenznummern der Eco-Version waren an die Modelle Monza Kaliber 12 angelehnt – 110.511 für PVD und 110.501 für das Chromgehäuse. Das Zifferblatt unterschied sich in Typo und Schriftstärke. An den Stunden-Indizes wurden neue rote Markierungen hinzugefügt, der weiße Datumsrahmen der früheren Versionen wurde weggelassen. Heuer verbaute bei der Eco-Version Kaliber 15-Gehäuse aus dem Lager mit den Referenzen 150.501/11. Diese Variante wurde bis Mitte der 80er Jahre gebaut und bildete den Abschluss der legendären Modellreihe Monza.

The clear, easy-to-read dial was framed by both a pulsometer and tachymeter scale—two essential tools for motorsport timing. The pulsometer allowed the user to measure heart rate by timing pulse beats, while the tachymeter scale enabled average speed to be calculated over a distance, based on the time taken.
The Monza's case back was engraved with the name "MONZA" above the Heuer logo, with reference and serial numbers typically located between the lugs at 6 and 12 o'clock. While the steel case back displayed the numbers clearly, the black and chrome-coated case versions could not show the engraving due to their finish. The crown on these models was unsigned.
In 1976, Heuer produced a promotional brochure for the Monza with references 150.501 (black PVD case) and 150.511 (chrome case). Featuring a photograph of Lauda wearing the watch, the brochure left no doubt that the Monza was the World Champion's watch. Heuer's sponsorship of Ferrari and its success in motorsport were explicitly highlighted, with the Monza model line praised for its groundbreaking design, quality, and affordability, aimed at a young, sporty demographic. The model line included:

1. *Monza Sport: black-coated with perforated leather strap*
2. *Lady Monza: black-coated with elegant black-and-white leather strap for women*
3. *Monza City: polished chrome case with stainless steel bracelet*

At the time, Heuer also offered a one-year warranty and the option to purchase the watches on an installment plan.
In the following year, Heuer expanded the Monza portfolio with a Calibre 12 version, reference 110.511.
The Monza series featured various case designs with differences in shape, coating, and size. Brass served as the base material for both the chrome and black PVD cases, with the black-coated versions available with or without brushed finish. The back covers varied in both engraving and typography.
Research suggests that the Monza series was discontinued from Heuer's catalogues after 1979, only to return in 1981 as the "Eco" version. This "Eco" version omitted the "Monza" name from the dial, used Calibre 12, and featured slight changes to the typography and font size. New red markings were added to the hour markers, and the white date frame from earlier versions was removed. Stock Calibre 15 cases (references 150.501/11) were used for this Eco version, which continued production into the mid-1980s, marking the end of the legendary Heuer Monza series.

Heuer verschenkt den Weltmeister.

Der Formel-1-Weltmeister trägt Heuer auf seinem Rennanzug, auf seinem Boliden und am Handgelenk. Darum möchte Ihnen Heuer eine kleine Freude machen.
Bei jedem sportlichen Uhrenfachhändler können Sie jetzt den Heuer-Gutschein einlösen. Und bei der Niki-Lauda-Show vom 27. 11. bis 5. 12. 1976 in Wien. Und vom 11. 12. bis 19. 12. 1976 in Salzburg.

Sportex Junior. Jugendstoppuhr in zwei Ausführungen: 1/5-Sekunden-Messer mit 0- bis 10-Minutenzähler und 1/10-Sekunden-Messer mit 0- bis 5-Minutenzähler. Farbige Drücker, griffsicheres Gehäuse rot oder schwarz. Eingebaute Halskordel. Preis: öS 395,–

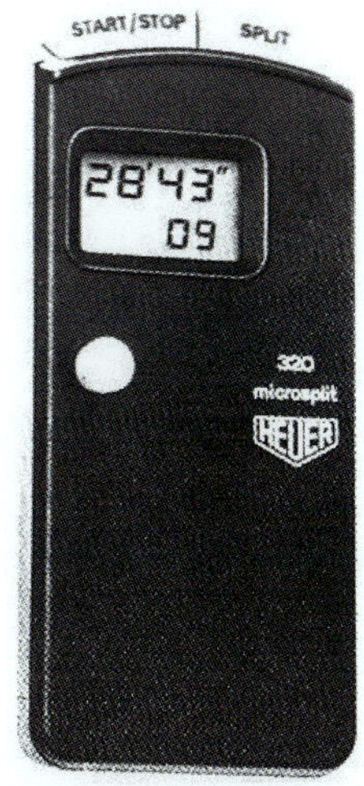

Microsplit 320. Die kleinste Quarzstoppuhr der Welt. Flüssigkristallanzeige (LCD) für perfekte Lesbarkeit im hellen Tageslicht. Zwischenzeitspeicher, Meßgenauigkeit: 1/100 Sekunde. Kompakte, handliche Form. Preis: öS 2.550,–

Monza. Automatischer Chronograph mit mattschwarzem Finish. Klares, sachliches Zifferblatt mit Tachymeter- und Pulsometerskala. Datum. Preis: öS 2.995,–

Chronosplit. Vollelektronischer Chronograph mit zwei getrennten Anzeigen: LCD-Anzeige für Tageszeit und Datum, LED-Anzeige auf die 1/10-Sekunde genau für Stoppuhr mit Zwischenzeitspeicher. Preis: öS 12.000,–

LCD Lady. Die solid-state Uhr für sportliche Damen. Flüssigkristallanzeige (LCD) mit Nachtbeleuchtung. Die einzige Damenuhr von Heuer. Preis: öS 3.950,–

Unverbindlich empfohlene Richtpreise.

Unter diesem Heuer-Autokleber finden Sie den Heuer-Gutschein.

CHRONOGRAPH
HEUER

Der Heuer-Hit des Jahres.
Cavallino. Tischuhr in originalgetreuer Nachbildung des Rennhelmes von Niki Lauda, Clay Regazzoni, Jacky Ickx, Johnny Cecotto. Neutrale De-Luxe-Ausführung in Chrom. Das ideale Geschenk für den Motorsportfan. Preis: öS 675,–

MONZA
HEUER
TACHYMETER
PULSOMETER
MONZA
HEUER
AUTOMATIC
CHRONOGRAPH
SWISS

MONZA

ref. 150.501 „Monza Sport“

Schwarz beschichtetes Gehäuse mit verschraubtem Boden. Das Zifferblatt ist schwarz mit einer kleinen permanenten Sekunde bei 10 Uhr und einem 30-Minuten-Hilfszifferblatt mit weißem Zeiger und roten gebogenen Balken. Das Zifferblatt ist von einem schwarzen Tachymeter-Ring umgeben , zwischen 12 und 3 Uhr ist ein rotes Pulsometer aufgedruckt. „MONZA“ steht über dem Heuer-Logo unter 12 Uhr und „AUTOMATIC CHRONOGRAPH“ über dem Datumsfenster bei 6 Uhr. Die mit Leuchtmasse gefüllten Stundenindizes sind gleichmäßig vergilbt.
Weiß bemalte Stahlzeiger haben ein schwarzes Zentrum und Leuchtmasse-Einlagen. Der rote Chronographen-Sekundenzeiger berührt die 1/5-Sekunden-Skala am äußeren Ende des Zifferblattes. Die Krone ist links und unsigniert. Die Drücker sind rechts und auf der Rückseite sichtbar. Der Boden ist mit „MONZA“ über dem Heuer-Logo signiert, darunter steht die Referenz- und Seriennummer. Auf einem kreisförmigen Streifen rund um den Rand des Gehäuses ist „WATER-RESISTENT TESTED 3 ATM“, „STAINLESS STEEL BACK“ und „SWISS MADE“ eingraviert. Unter dem Heuer-Sticker zeigt der Boden einen Längsschliff.
Das Uhrwerk ist ein Kaliber 15 und mit „CAL. 15“ auf der Brücke signiert.

Black-coated automatic Monza screw-back case. The dial is black with a small permanent red second hand at 10 o'clock and a 30-minute recording subdial with a white hand and curved red bars. The dial is surrounded by a black tachymeter ring with an additional red printed pulsometer between 12 and 3 o'clock. Below 12 o'clock, "MONZA" is over the Heuer logo and "AUTOMATIC CHRONOGRAPH" above the date window at 6 o'clock. The luminous hour markers have all aged to a uniform yellow color, markers have been omitted at 3 and 6 o'clock.
The steel hands are painted white with a black center and luminous inserts. The red chronograph hand reaches the 1/5-second track on the outer edge of the dial. The crown is on the left side and unsigned. The pushers are on the right side of the case and visible from the back. The case back is signed "MONZA" above the Heuer logo, beneath the reference and serial number. Engraved in a circular path around the edge of the case is "WATER-RESISTANT TESTED 3 ATM," "STAINLESS STEEL BACK," and "SWISS MADE." The case back has longitudinal grinding under the original Heuer sticker.
The movement is a Calibre 15, and is signed "CAL. 15" on the bridge.

MONZA
HEUER
AUTOMATIC
CHRONOGRAPH
PULSOMETER
TACHYMETER
SWISS
15
ER Uhre
tät. Ihre G
Zuverlässigkeit
nwörtlich. HEUER is
der berühmtesten S

MONZA

ref. 150.501 „Lady Monza“

Schwarz überzogenes Gehäuse mit verschraubtem Boden. Das Zifferblatt ist schwarz mit einer kleinen permanenten Sekunde bei 10 Uhr und einem 30-Minuten-Hilfszifferblatt mit weißem Zeiger und roten gebogenen Balken. Das Zifferblatt ist von einem schwarzen Tachymeter-Ring umgeben, zwischen 12 und 3 Uhr ist ein rotes Pulsometer aufgedruckt. „MONZA“ steht über dem Heuer-Logo unter 12 Uhr und „AUTOMATIC CHRONOGRAPH“ über dem Datumsfenster bei 6 Uhr. Die mit Leuchtmasse gefüllten Stundenindizes sind gleichmäßig vergilbt.
Weiß bemalte Stahlzeiger haben ein schwarzes Zentrum und Leuchtmasse-Einlagen. Der rote Chronographen-Sekundenzeiger berührt die 1/5-Sekunden-Skala am äußeren Ende des Zifferblattes. Die Krone ist links und unsigniert. Die Drücker sind rechts und auf der Rückseite sichtbar. Der Boden ist mit „MONZA“ über dem Heuer-Logo signiert, darunter steht die Referenz- und Seriennummer. Auf einem kreisförmigen Streifen rund um den Rand des Gehäuses ist „WATER-RESISTENT TESTED 3 ATM“, „STAINLESS STEEL BACK“ und „SWISS MADE“ eingraviert. Unter dem Heuer-Sticker zeigt der Boden einen Längsschliff.
Das Uhrwerk ist ein Kaliber 15 und mit „CAL. 15“ auf der Brücke signiert.

Black-coated automatic Monza screw-back case. The dial is black with a small permanent red second hand at 10 o'clock and a 30-minute recording subdial with a white hand and curved red bars. The dial is surrounded by a black tachymeter ring with an additional red printed pulsometer between 12 and 3 o'clock. Below 12 o'clock, "MONZA" is over the Heuer logo and "AUTOMATIC CHRONOGRAPH" above the date window at 6 o'clock. The luminous hour markers have all aged to a uniform yellow color; markers have been omitted at 3 and 6 o'clock.
The steel hands are painted white with a black center and luminous inserts. The red chronograph hand reaches the 1/5-second track on the outer edge of the dial. The crown is on the left and unsigned. The pushers are on the right of the case and visible from the back. The case back is signed "MONZA" above the Heuer logo, beneath the reference and serial number. Engraved in a circular path around the edge of the case is "WATER-RESISTANT TESTED 3 ATM," "STAINLESS STEEL BACK," and "SWISS MADE." The case back has longitudinal grinding under the original Heuer sticker.
The movement is a Calibre 15, and is signed "CAL. 15" on the bridge.

MONZA

ref. 150.511 „Monza City“

Verchromtes Gehäuse mit verschraubtem Boden, oben und seitlich mit Strichschliff. Das Zifferblatt ist schwarz mit einer kleinen permanenten Sekunde bei 10 Uhr und einem 30-Minuten-Hilfszifferblatt mit weißem Zeiger und roten gebogenen Balken. Das Zifferblatt ist von einem schwarzen Tachymeter-Ring umgeben, zwischen 12 und 3 Uhr ist ein rotes Pulsometer aufgedruckt. „MONZA“ steht über dem Heuer-Logo unter 12 Uhr und „AUTOMATIC CHRONOGRAPH“ über dem Datumsfenster bei 6 Uhr. Die mit Leuchtmasse gefüllten Stundenindizes sind gleichmäßig vergilbt. Keine Indizes findet man bei 3 und 6 Uhr.
Weiß bemalte Stahlzeiger haben ein schwarzes Zentrum und Leuchtmasse-Einlagen. Der rote Chronographen-Sekundenzeiger berührt die 1/5-Sekunden-Skala. Die Krone ist links und unsigniert. Die Drücker sind rechts und auf der Rückseite sichtbar. Der Rückdeckel ist mit „MONZA“ über dem Heuer-Logo signiert, darunter steht die Referenz- und Seriennummer. Auf einem kreisförmigen Streifen rund um den Rand des Gehäuses ist „WATER-RESISTENT TESTED 3 ATM“, „STAINLESS STEEL BACK“ und „SWISS MADE“ eingraviert. Unter dem Original-Siegel zeigt der Boden einen Längsschliff. Das Uhrwerk ist ein Kaliber 15 und mit „CAL. 15“ auf der Brücke signiert.

Chrome-plated screw-back case with brushed finish on top and sides. The dial is matt black with a small permanent red second hand at 10 o'clock and a 30-minute recording subdial with a white hand and curved red bars.The dial is surrounded by a black tachymeter ring with an additional red printed pulsometer between 12 and 3 o'clock. Underneath 12 o'clock, "MONZA" is over the Heuer logo and "AUTOMATIC CHRONOGRAPH" above the date window at 6 o'clock. The luminous hour markers have all aged to a uniform yellow color; markers have been omitted at 3 and 6 o'clock.
The steel hands are painted white with a black center and luminous inserts. The red chronograph hand reaches the 1/5-second track on the outer edge of the dial. The crown is on the left and unsigned. Chronograph pushers are integrated on the right side of the case fully visible from the back. The case back is signed "MONZA" above the Heuer logo, beneath the reference and serial number. Engraved in a circular path around the edge of the case are "WATER-RESISTANT TESTED 3 ATM," "STAINLESS STEEL BACK," and "SWISS MADE." The case back has longitudinal grinding under the original seal.
The movement is a Calibre 15, and is signed "CAL. 15" on the bridge.

HEUER
MONZA
HEUER
AUTOMATIC
CHRONOGRAPH

Werbebroschüre für die HEUER Monza aus dem Jahr 1976.
HEUER Monza leaflet from 1976.

Überall, wo Zeitnahme, Leistung und Präzision zählen, verlassen sich Männer wie ich auf Spezialuhren von Heuer. Heuer-Chronographen gelten als die Uhren der internationalen Sportelite wegen ihrer serienerprobten Zuverlässigkeit und sichtbaren sportlichen Eleganz. Mit dem neuen Monza-Automatic-Chronographen – erschienen aus Anlaß der Automobil-Weltmeisterschaft 1975/76 – ist Heuer eine weitere Spitzenleistung gelungen. Überzeugen Sie sich selbst.

Ihr

Niki Lauda

Die Entscheidung fiel in Monza: Niki Lauda wurde Weltmeister auf Ferrari.

Eindrucksvollster Beweis für den hohen Qualitätsstandard der HEUER-Produkte:

HEUER

ist der offizielle Zeitnehmer von FERRARI

und kann auf eine langjährige und erfolgreiche Zusammenarbeit mit dem berühmten italienischen Sport- und Rennwagen-Hersteller zurückblicken. Ausgerüstet mit komplizierten elektronischen Meßanlagen begleiten erfahrene HEUER-Ingenieure das Team von FERRARI an alle Rennstrecken der Welt.

MONZA-TESTGUTSCHEIN NOCH HEUTE ABSCHICKEN!

✱ Der HEUER-MONZA-AUTOMATIC-CHRONOGRAPH erscheint exklusiv zur Automobil-Weltmeisterschaft 1975/76: Niki Lauda siegte auf FERRARI!

MONZA

ref. 110.511

Schwarz beschichtetes Gehäuse mit verschraubtem Boden, oben und seitlich mit Strichschliff. Das Zifferblatt ist schwarz mit einem 30 Minuten sowie 12 Stunden Hilfszifferblatt und weißen rechteckigen Zeigern.
Abgerundet wird das Blatt von einem schwarzen Tachymeter-Ring mit roter Pulsometer-Skala zwischen 12 und 3 Uhr. Der Modellname „MONZA" steht über dem Heuer Logo unter 12 Uhr sowie „AUTOMATIC CHRONOGRAPH" über dem mit „SWISS" signierten und weiß umrandeten Datumsfenster. Die mit Leuchtmasse gefüllten Stundenindizes sind ebenmäßig gelblich patiniert. Keine Indizes findet man bei 3, 6 und 9 Uhr.
Weiße Stahlzeiger haben eine schwarze Mitte und längliche Leuchtmasse-Einschlüsse. Der angespitzte rote Chronographen-Sekundenzeiger berührt präzise die 1/5-Sekunden-Skala am äußeren Ende des Zifferblattes. Die Krone ist links und signiert. Die Chronographen-Drücker sind rechts integriert und von hinten sichtbar. Der Rückdeckel ist mit „MONZA" signiert über dem Heuer-Logo, darunter steht die Referenz- und Seriennummer. Auf einem kreisförmigen Streifen rund um den Rand des Gehäuses ist „WATER- RESISTENT TESTED 3 ATM", „STAINLESS STEEL BACK" und „SWISS MADE" eingraviert. Der Rückdeckel zeigt einen Längsschliff unter dem Originalsiegel.
Das Uhrwerk ist ein Kaliber 12 und auf der Brücke CAL. 12 signiert.

Black-coated automatic Monza screw-back case with brushed finish on top and sides. The dial is black with 30-minute and 12-hour recording subdials and white hands. Three curved red bars divide the minutes subdial.
The dial is surrounded by a black tachymeter ring with an additional red printed pulsometer between 12 and 3 o'clock. Underneath 12 o'clock, "MONZA" is above the Heuer logo. "AUTOMATIC CHRONOGRAPH" is above the "SWISS"-signed and white-framed date window at 6 o'clock. The luminous hour markers have all aged to a uniform yellow color, markers have been omitted at 3, 6, and 9 o'clock. Steel hands are painted white with a black centre and luminous inserts. A tipped red chronograph hand reaches the 1/5-second-track on the outer edge of the dial precisely. The crown on the left is signed "HEUER." Chronograph pushers are integrated on the right of the case, fully visible from the back. The case back is signed "MONZA" over the Heuer logo beneath the reference and serial number. Engraved in a circular path around the edge of the case are "WATER-RESISTANT TESTED 3 ATM," "STAINLESS STEEL BACK," and "SWISS MADE." The case back has longitudinal grinding under the original seal. The movement is a Calibre 12 signed "CAL.12" on the bridge.

TACHYMETER
PULSOMETER
MONZA
HEUER
AUTOMATIC
CHRONOGRAPH
SWISS

MONZA

ref. 110.501 „Economy“

Verchromtes Gehäuse mit verschraubtem Boden oben und seitlich mit Strichschliff. Das Zifferblatt ist schwarz mit einem 30-Minuten- sowie 12-Stunden-Hilfszifferblatt und weißen rechteckigen Zeigern.
Abgerundet wird das Blatt von einem schwarzen Tachymeter-Ring mit roter Pulsometer-Skala zwischen 12 und 3 Uhr. Der Modellname „MONZA“ verschwand bei der „Eco“-Version, das Heuer-Logo stand unter 12 Uhr sowie „AUTOMATIC CHRONOGRAPH“ über dem Datum bei 6 Uhr. Die übliche weiße Umrandung des Datumsfensters fehlte. Die mit Leuchtmasse gefüllten Stundenindizes sind ebenmäßig gelblich patiniert, mit roten Strichen an der Stundenskalierung. Keine Indizes findet man bei 3, 6 und 9 Uhr.
Weiße Stahlzeiger haben eine schwarze Mitte und längliche Leuchtmasse-Einschlüsse. Der angespitzte rote Chronographen-Sekundenzeiger berührt präzise die 1/5-Sekunden-Skala am äußeren Ende des Zifferblattes. Die Krone ist links und unsigniert. Die Chronographen-Drücker sind rechts integriert und von hinten sichtbar. Der Rückdeckel ist mit „MONZA“ signiert über dem Heuer-Logo, darunter steht die Referenz- und Seriennummer. Auf einem kreisförmigen Streifen rund um den Rand des Gehäuses ist „WATER- RESISTENT TESTED 3 ATM“, „STAINLESS STEEL BACK“ und „SWISS MADE“ eingraviert. Der Rückdeckel zeigt einen Längsschliff.
Das Uhrwerk ist ein Kaliber 12 und mit „CAL. 12“ auf der Brücke signiert.

Chrome-plated automatic Monza screw-back case with brushed finish on top and sides. The dial is black with 30-minute and 12-hour recording subdials with white hands. Three curved red bars divide the minute subdial.
The dial is surrounded by a black tachymeter ring with an additional red printed pulsometer between 12 and 3 o'clock. Underneath 12 o'clock, the "MONZA" name above the Heuer logo in the Eco version. "AUTOMATIC CHRONOGRAPH" remains above "SWISS," with a full black date window without white frame at 6 o'clock. The luminous hour markers with red accents have aged to a uniform yellow color; markers have been omitted at 3, 6, and 9 o'clock. The steel hands are painted white with a black center and luminous inserts. A tipped red chronograph hand reaches the 1/5-second track on the outer edge of the dial precisely. The crown on the left is unsigned. Chronograph pushers are integrated on the right of the case, fully visible from the back. The case back is signed "MONZA" over the Heuer logo beneath the reference and serial number. Engraved in a circular path around the edge of the case can be seen "WATER-RESISTANT TESTED 3 ATM," "STAINLESS STEEL BACK," and "SWISS MADE." The case back has longitudinal grinding.
The movement is a Calibre 12 signed "CAL.12" on the bridge.

TACHYMETER
PULSOMETER
HEUER
AUTOMATIC
CHRONOGRAPH
SWISS

HEUER
HEUER
MODENA
HEUER
AUTOMATIC
CHRONOGRAPH

MODENA

Modena ist eine italienische Stadt in der Emilia Romagna mit 180.000 Einwohnern. Einerseits ist sie eine beliebte Universitätsstadt, anderseits international bekannt für ihre Automobilindustrie. So liegt die Fabrik von Maserati direkt in Modena und auch Lamborghini sitzt nur einen Steinwurf entfernt in Sant'Agata Bolognese.
Einer der berühmtesten Bürger der Stadt war Enzo Anselmo Ferrari, der 1898 in Monza das Licht der Welt erblickte und 90 Jahre später dort verstarb.
Bereits 1929 gründete der junge Ferrari seinen eigenen Rennstall in Modena. Nachdem seine Werkstatt im Zweiten Weltkrieg zerstört wurde, zog Ferrari mit seinem Rennstall ins nur 20 km entfernte Maranello.
Ferraris Geburtshaus in Modena beherbergt heute ein großes Museum. „Il Commentatore", wie Ferrari oft genannt wurde, blieb seiner Stadt ein Leben lang treu.
Jack Heuer hatte einen guten Draht zu Enzo Ferrari, den er als harten Verhandler beschrieb. Mit Heuer als Sponsor geriet die Scuderia Ferrari auf steten Erfolgskurs, der in drei Formel 1- Weltmeistertiteln (1975, 1977, 1979) gipfelte.
Jack Heuer hatte generell eine Schwäche für italienische Namen: „Weil sie so gut klingen, Herr Haslinger", sagte er mir einst in einem Interview. Daher erschien nach dem zweiten Weltmeistertitel, wieder durch Niki Lauda errungen, in einer sehr begrenzten Stückzahl, eine Sonderedition mit dem klingenden Namen: Heuer Modena!
Sie entsprach technisch wie optisch der Monza, die Rückdeckel trugen sogar die Monza-Signatur sowie identische Referenznummern.
Das Gehäuse der Modena Kaliber 15 mit Streifenschliff ist wuchtiger, die Variante Kaliber 12 mit glatter Oberfläche ist schlanker. Die Zifferblätter variierten in Typo, Größe und Farbe des Druckes. Der rote Chronographen-Zeiger berührt mit seiner Spitze millimetergenau die 1/5-Sekunden-Skala für perfekte Ablesbarkeit. Bei beiden Modenas war die Krone mit „Heuer" signiert.
Die Modena, mit ihrer eng mit Ferrari verwobenen Geschichte, ist mit Sicherheit eines der seltensten Modelle und bei Sammlern sehr begehrt.

Modena is an Italian city with a population of 180,000, located in the Emilia-Romagna region. It is a thriving university town and internationally renowned for its automotive industry. Maserati operates its factory directly in Modena, while Lamborghini's headquarters are just a stone's throw away in Sant'Agata Bolognese.
One of the city's most famous residents was Enzo Anselmo Ferrari, who was born in Modena in 1898 and died there 90 years later.
Ferrari founded his own racing team in Modena as early as 1929. After his workshop was destroyed during the Second World War, he moved his racing team to Maranello, just 20 kilometers (12.4 miles) away.
Ferrari's birthplace in Modena is now home to a large museum. "Il Commentatore," as Ferrari was often called, remained loyal to his hometown throughout his life. Jack Heuer had a strong relationship with Enzo Ferrari, describing him as a tough negotiator. With Heuer as a sponsor, Scuderia Ferrari's fortunes flourished, culminating in three Formula 1 World Championship titles in 1975, 1977, and 1979. Heuer was particularly fond of Italian names: 'Because they sound so good, Mr. Haslinger,' he once told me in an interview. After Ferrari's second world championship win, again with Niki Lauda as the driver, Heuer released a special edition watch in very limited numbers, named the Heuer Modena.
The Modena closely resembled the Monza, both technically and visually. Even the case backs bore the Monza signature, as well as identical reference numbers.
The Modena with Calibre 15 features a bulkier case with a stripe finish, while the Calibre 12 version has a slimmer case with a smooth surface. The dials vary in typography, size, and print color. The red chronograph hand precisely aligns with the 1/5-second scale, ensuring perfect readability. Both Modena models feature a Heuer-signed crown.
With its strong connection to Ferrari's legacy, the Heuer Modena is one of the rarest and most sought-after models among collectors.

MODENA

ref. 150.501

Schwarz beschichtetes Modena-Gehäuse mit verschraubtem Boden, leichtem Bürstenschliff oben und seitlich. Das schwarze Zifferblatt besteht aus einer kleinen permanenten Sekunde bei 10 Uhr sowie einem 30-Minuten-Hilfszifferblatt mit weißem Zeiger über roten fünf Minuten Balken.
Das Zifferblatt wird von einem schwarzen Tachymeter-Ring abgerundet, mit einer roten Pulsometer-Skala zwischen 12 und 3 Uhr. „MODENA“ steht über dem Heuer-Logo unter 12 Uhr und „AUTOMATIC CHRONOGRAPH“ über dem weiß umrandeten Datumsfenster bei 6 Uhr. Die mit Leuchtmasse gefüllten Stundenindizes sind gleichmäßig vergilbt. Keine Indizes findet man bei 3 und 6 Uhr.
Weiße Stahlzeiger haben eine schwarze Mitte und längliche Leuchtmasse-Einschlüsse. Der rote angespitzte Chronographen-Sekundenzeiger berührt präzise die 1/5-Sekunden-Skala am äußeren Ende des Zifferblattes. Die Krone ist links und mit „HEUER“ signiert. Die Chronographen-Drücker sind rechts integriert und von hinten vollkommen sichtbar. Der Rückdeckel ist mit „MONZA“ signiert über dem Heuer-Logo, darunter steht die Referenz- und Seriennummer. Auf einem kreisförmigen Streifen rund um den Rand des Gehäuses ist „WATER- RESISTENT TESTED 3 ATM“, „STAINLESS STEEL BACK“ und „SWISS MADE“ eingraviert. Der Rückdeckel zeigt einen Längsschliff.
Das Uhrwerk ist ein Caliber 15 und mit „CAL. 15“ auf der Brücke signiert.

Black-coated Modena screw-back case with brushed finish on top and sides. The dial is black with a small permanent red second hand at 10 o'clock, a 30-minute recording subdial, a white hand and curved red bars.
The dial is surrounded by a black tachymeter ring with an additional red printed pulsometer between 12 and 3 o'clock. Underneath 12 o'clock, "MODENA" is over the Heuer logo and "AUTOMATIC CHRONOGRAPH" above the white-framed date window at 6 o'clock. The luminous hour indices have aged to a uniform yellow color; markers have been omitted at 3 and 6 o'clock. The steel hands are painted white with a black center and luminous inserts. A tipped red chronograph hand reaches the 1/5-second track on the outer edge of the dial. The crown on the left is signed "HEUER." Chronograph pushers are integrated on the right of the case, fully visible from the back.
The case back is signed "MONZA" over the Heuer logo beneath the reference and serial number. Engraved in a circular path around the edge of the case are "WATER-RESISTANT TESTED 3 ATM," "STAINLESS STEEL BACK," and "SWISS MADE." The case back has longitudinal grinding.
The movement is a Calibre 15 signed "CAL.15" on the bridge.

TACHYMETER
PULSOMETER
MODENA
HEUER
AUTOMATIC
CHRONOGRAPH
SWISS
6

MODENA

ref. 110.511

Schwarz überzogenes Modena-Gehäuse mit verschraubtem Boden. Das Zifferblatt ist schwarz mit einem 30-Minuten- sowie 12-Stunden-Hilfszifferblatt und weißen rechteckigen Zeigern. Abgerundet wird das Blatt von einem schwarzen Tachymeter-Ring mit roter Pulsometer-Skala zwischen 12 und 3 Uhr.
Der Modelname „MODENA“ steht über dem Heuer-Logo unter 12 Uhr sowie „AUTOMATIC CHRONOGRAPH“ über dem mit „SWISS“ signierten und weiß umrandeten Datumsfenster. Leuchtmasse gefüllte Stundenindizes sind ebenmäßig gelblich patiniert. Keine Indizes findet man bei 3, 6 und 9 Uhr.
Weiße Stahlzeiger haben eine schwarze Mitte und längliche Leuchtmasse-Einschlüsse. Der rote Chronographen-Sekundenzeiger berührt präzise die 1/5-Sekunden-Skala am äußeren Ende des Zifferblattes. Die Krone ist links und mit „HEUER“ signiert. Die Chronographen-Drücker sind rechts integriert und von hinten vollkommen sichtbar. Der Rückdeckel ist mit „MONZA“ signiert über dem Heuer-Logo, darunter steht die Referenz- und Seriennummer. Auf einem kreisförmigen Streifen rund um den Rand des Gehäuses ist „WATER- RESISTENT TESTED 3 ATM“, „STAINLESS STEEL BACK“ und „SWISS MADE“ eingraviert. Unter dem Original-Siegel zeigt der Rückdeckel einen Längsschliff.
Das Uhrwerk ist ein Kaliber 12 und auf der Brücke CAL. 12 signiert.

Black-coated automatic Modena screw-back case. The dial is black with 30-minute and 12-hour recording subdials and white hands. Three curved red bars divide the minute subdial.
The dial is surrounded by a black tachymeter ring with an additional red printed pulsometer between 12 and 3 o'clock. Underneath 12 o'clock, "MODENA" is above the Heuer logo. "AUTOMATIC CHRONOGRAPH" is above the "SWISS"-signed and white-framed date window at 6 o'clock. The luminous hour markers have all aged to a uniform yellow color; markers have been omitted at 3, 6, and 9 o'clock.The steel hands are painted white with a black center and luminous inserts. A tipped red chronograph hand reaches the 1/5-second track on the outer edge of the dial precisely. The crown on the left is signed "HEUER." Chronograph pushers are integrated on the right of the case, fully visible from the back. The case back is signed "MONZA" over the Heuer logo beneath the reference and serial number. Engraved in a circular path around the edge of the case can be seen "WATER-RESISTANT TESTED 3 ATM," "STAINLESS STEEL BACK," and "SWISS MADE." The case back has longitudinal grinding under the original seal.
The movement is a Calibre 12 signed "CAL.12" on the bridge.

TACHYMETER
PULSOMETER
MODENA
HEUER
AUTOMATIC
CHRONOGRAPH
SWISS

HAILWOOD
MIKE HAILWOOD AND TED MACAULEY
T.MACAULEY
TEAM HAILWOOD
IOM TT
1978

SILVERSTONE

Silverstone ist ein kleiner Ort zwischen Northhampton und Oxford, ungefähr eine Fahrstunde westlich von London. Seit 1948 werden hier auf einem ehemaligen Flughafen Autorennen gefahren. Für Ferrari-Fans ist dies eine ganz besondere Rennstrecke, da José Froilán González hier 1951 den ersten Sieg für die Scuderia einfahren konnte. Der Hochgeschwindigkeitskurs bot viele unglaublich spannende Rennen, auch blieb die Streckenführung über fast 30 Jahre unverändert. Ein fantastisches Bild von Clay Regazzoni erscheint im Heuer-Katalog von 1974, es zeigt ihn im blauen Rennanzug im Cockpit seines Ferrari sitzend, am linken Arm glänzt die blaue Silverstone mit Stahlband. Eine glaubwürdige Darstellung der gefährlichen Hochgeschwindigkeits-Welt der damaligen Formel 1 - und ein wirkungsvolles Image für den Sponsor.

Die Heuer Silverstone wird 1974 lanciert. Das Gehäuse-Design ist absolut neuartig, eine kreative Meisterleistung, rechteckig mit abgerundeten Kanten und hochglänzend poliert. Auf der Rückseite ist ein Längsschliff angebracht, die Krone wird durch eine leichte Einkerbung an der Seite geschützt.

Als Antrieb diente das bewährte Kaliber 12 Automatik-Werk mit Mikrorotor. Die beiden Drücker befinden sich daher gegenüber der Krone auf der rechten Seite in Aussparungen.

Das Zifferblatt hat eine 30-Minuten- und 12-Stunden-Anzeige und das Datum bei 6 Uhr. Neben dem blauen Zifferblatt gab es noch Varianten in Bordeauxrot und in Braun. Am Blattrand beginnt eine innen liegende Lünette, die schräg zum Gehäuserand ansteigt, eine 1/5-Sekunden-Skala und rechteckige Leuchtpunkte beinhaltet. Die Heuer Silverstone wurde bis 1977 immer wieder in den Heuer-Katalogen gezeigt.

Silverstone is a small village an hour's drive west of London, between Northampton and Oxford. Car races have been held on the grounds of the former airport here since 1948.

As the location of the Scuderia's first victory in 1951 with Jose Froilan Gonzalez behind the wheel, it is a special place for Ferrari fans. This high-speed race track provided for enormously exciting races. The layout of the track remained the same for almost thirty years.

A fantastic photograph of Clay Regazzoni appeared in the Heuer catalogue in 1974. It shows him sitting in the cockpit of his Ferrari in a blue racing suit-wearing a glittering blue Heuer Silverstone with a glittering steel band on his left arm. This portrayal impressively communicates the risk-laden high-speed world of Formula One and was an effective image for sponsors.

The Heuer Silverstone was launched in 1974. The case design was completely new, a creative square masterpiece with highly polished rounded edges. The backing had longitudinal grinding; the crown was protected by a slight indentation on the side.

It was driven by the proven Calibre 12 automatic movement with a microrotor. The two pushers were placed across from the crown in two indentations. The dial had a 30-minute and 12-hour display, and the date was located at 6 o'clock. As well as the blue dial, other variations were burgundy and brown. At the edge of the dial an internal bezel sloped up to the case edge, a 1/5-second scale, and rectangular luminescent points.

The Heuer Silverstone was presented in the Heuer catalogues until 1977.

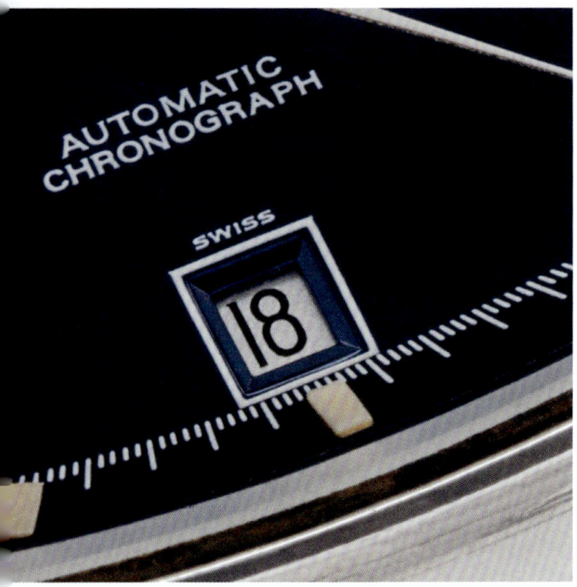

SILVERSTONE

ref. 110.313 B

Quadratisches Gehäuse mit gerundeten Ecken, verschraubtem Boden und polierter Oberfläche sowie polierten Seiten. Glänzend tiefblaues Zifferblatt mit 30-Minuten- und 12-Stunden-Hilfszifferblättern, umgeben von einem angewinkelten großen Armierungsring in gleicher Farbe.
Der Modellname „SILVERSTONE“ steht unter dem Heuer-Logo unter 12 Uhr. „AUTOMATIC CHRONOGRAPH“ steht über dem weiß gerahmten Datumsfenster bei 6 Uhr. Das Datum ist schwarz auf silbernem Untergrund. Stundenindizes mit Leuchtmasse sind am äußeren Ende des Spannungsrings neben der 1/5-Sekunden-Skala angebracht. Das 30-Minuten-Hilfszifferblatt zeigt die Zahlen 5, 10, 15, 20, 25 und 30 anstelle der Standard-Konfiguration von 10, 20 und 30. Die weiß bemalten Zeiger haben Leuchtmasse-Einlagen. Der weiße Chronographen-Sekundenzeiger zeigt auf die weiße 1/5-Sekunden-Skala am äußeren Rand des Zifferblattes. Die Krone ist links integriert und mit „HEUER“ signiert. Die Drücker sind rechts integriert.
Der Gehäuseboden hat einen schönen Bürstenschliff und der Rückdeckel einen Kreisschliff mit polierten Seiten.
Das Uhrwerk ist ein Kaliber 12 und mit „CAL. 12“ auf der Brücke signiert.

Square screw-back case with rounded edges and polished top and sides. The gleaming deep blue dial has 30-minute and 12-hour recording subdials. These are surrounded by a large angled tension ring in the same color.
The model name "SILVERSTONE" is positioned under the Heuer logo below 12 o'clock. "AUTOMATIC CHRONOGRAPH" is above the white-framed date window at 6 o'clock. The date is in black on a silver background. Luminous hour markers are placed on the outer edge of the tension ring beside the second track. The minute subdial shows 5, 10, 15, 20, 25, and 30 instead of the standard 10, 20, and 30 configuration. The white-painted hands have large luminous inserts. The white chronograph second hand is aligned to the white 1/5-second track on the outer edge of the dial. The crown is integrated on the left of the case and signed "HEUER." Chronograph pushers are integrated on the right of the case.
The base has a beautiful brushed finish and the case back features circle grinding with polished sides.
The movement is a Calibre 12, and is signed "CAL. 12" on the bridge.

Heuer macht ke
gewöhnlichen U
Der Bewei
HEUER
SILVERSTONE
AUTOMATIC
CHRONOGRAPH
SWISS
Silverstone. Exklusiver automatischer Chronograph. Wasserdichtes Edelstahlgehäuse mit Mineralglas und Leuchtzifferblatt. Zeigt die Zeit und stoppt die Zeit. Keine gewöhnliche Uhr. Eine Heuer.

Clay Regazzoni trägt, im Ferrari sitzend, die blaue Silverstone, 1974.

Clay Regazzoni wearing the blue Silverstone sitting in the Ferrari in 1974.

SILVERSTONE

ref. 110.313 R

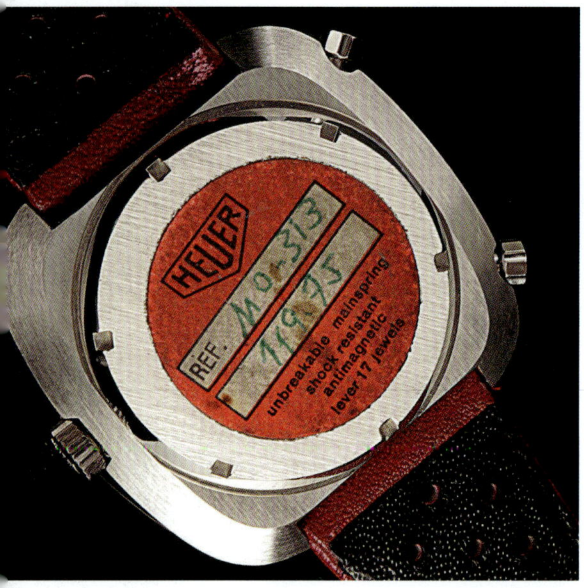

Quadratisches Gehäuse mit gerundeten Ecken, verschraubtem Boden und polierter Oberfläche sowie polierten Seiten. Bordeauxrotes Zifferblatt mit 30-Minuten- und 12-Stunden-Hilfszifferblättern, umgeben von einem angewinkelten großen Armierungsring in gleicher Farbe.
Der Modellname „SILVERSTONE“ steht unter dem Heuer-Logo unter 12 Uhr. „AUTOMATIC CHRONOGRAPH“ steht über dem weiß gerahmten Datumsfenster bei 6 Uhr. Das Datum ist schwarz auf silbernem Untergrund. Stundenindizes mit Leuchtmasse sind am äußeren Ende des Spannungsrings neben der 1/5-Sekunden-Skala angebracht. Das 30-Minuten-Hilfszifferblatt zeigt die Zahlen 5, 10, 15, 20, 25 und 30 anstelle der Standard-Konfiguration von 10, 20 und 30. Die weiß bemalten Zeiger haben Leuchtmasse-Einlagen. Der weiße Chronographen-Sekundenzeiger zeigt auf die weiße 1/5-Sekunden-Skala am äußeren Rand des Zifferblattes. Die Krone ist auf der linken Seite des Gehäuses integriert und mit „HEUER“ signiert. Die Chronographen-Drücker sind auf der rechten Seite integriert. Der Gehäuseboden hat einen schönen Bürstenschliff und der Rückdeckel einen Längsschliff mit polierten Seiten. Das Uhrwerk ist ein Kaliber 12 und mit „CAL. 12“ auf der Brücke signiert.

Square screw-back case with rounded edges and polished top and sides. The burgundy dial has 30-minute and 12-hour recording subdials and is surrounded by a large angled tension ring in the same color.
The model name "SILVERSTONE" is positioned under the Heuer logo below 12 o'clock. "AUTOMATIC CHRONOGRAPH" is above the white-framed date window at 6 o'clock. The date is in black on a silver background. Luminous hour markers are placed on the outer end of the tension ring beside the second track. The minute subdial shows 5, 10, 15, 20, 25, and 30 instead of the standard 10, 20, and 30 configuration.
The white-painted hands have large luminous inserts. The white chronograph second hand aligned to the white 1/5-second track on the outer edge of the dial. The crown is integrated on the left of the case and signed "HEUER." Chronograph pushers are integrated on the right of the case. The base has a beautiful brushed finish and the case back features longitudinal grinding with polished sides.
The movement is a Calibre 12, and is signed "CAL.12" on the bridge.

AUTOMATIC
CHRONOGRAPH

HEUER
SILVERSTONE
AUTOMATIC
CHRONOGRAPH
SWISS
18

SILVERSTONE

ref. 110.313 F

Quadratisches Gehäuse mit gerundeten Ecken, verschraubtem Boden und polierter Oberfläche sowie polierten Seiten. Braunes Zifferblatt mit Sternenschliff und 30-Minuten- und 12-Stunden-Hilfszifferblättern, umgeben von einem angewinkelten großen Armierungsring in gleicher Farbe. Der Modellname „SILVERSTONE“ steht unter dem Heuer-Logo unter 12 Uhr. „AUTOMATIC CHRONOGRAPH“ steht über dem weiß gerahmten Datumsfenster bei 6 Uhr. Das Datum ist schwarz auf silbernem Untergrund. Stundenindizes mit Leuchtmasse sind am äußeren Ende des Spannungsrings neben der 1/5-Sekunden-Skala angebracht. Das 30-Minuten-Hilfszifferblatt zeigt die Zahlen 5, 10, 15, 20, 25 und 30 anstelle der Standard-Konfiguration von 10, 20 und 30.
Die weiß bemalten Zeiger haben Leuchtmasse-Einlagen. Der weiße Chronographen-Sekundenzeiger zeigt auf die 1/5-Sekunden-Skala. Die Krone ist auf der linken Seite des Gehäuses integriert und mit „HEUER“ signiert. Die Chronographen-Drücker sind auf der rechten Seite integriert. Der Gehäuseboden hat einen schönen Bürstenschliff und der Rückdeckel einen Längsschliff mit polierten Seiten. Das Uhrwerk ist ein Kaliber 12 und mit „CAL. 12“ auf der Brücke signiert.

Square screw-back case with rounded edges and polished top and sides. The brown dial has star grinding and 30-minute and 12-hour recording subdials and is surrounded by a large angled tension ring in the same color.
The model name “SILVERSTONE” is positioned under the Heuer logo below 12 o’clock. “AUTOMATIC CHRONOGRAPH” is above the white-framed date window at 6 o’clock. The date is in black on a silver background. Luminous hour markers are placed on the outer edge of the tension ring beside the second track. The minute subdial shows 5, 10, 15, 20, 25, and 30 instead of the standard 10, 20, and 30 configuration.
The white-painted hands have large luminous inserts. The white chronograph second hand points at the white 1/5-second track on the outer edge of the dial. The crown is integrated on the left of the case and signed “HEUER.” Chronograph pushers are integrated on the right of the case.
The base has a beautiful brushed finish and the case back has longitudinal grinding with polished sides. The movement is a Calibre 12, and is signed “CAL.12” on the bridge.

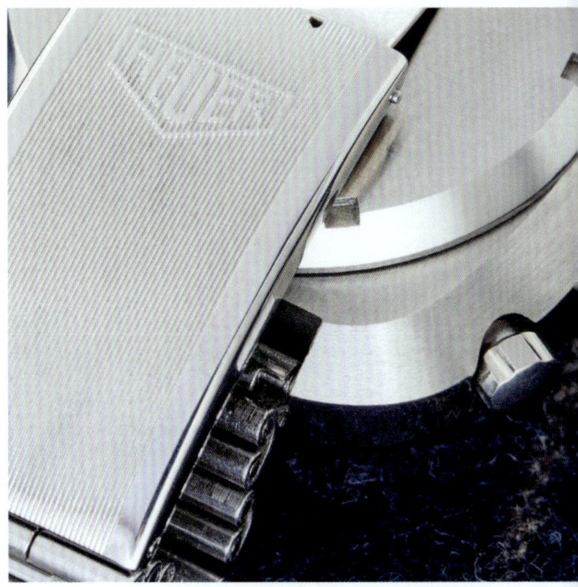

The Haslinger Collect
Wednesday 15 Dece
New Bond Street
Selected highli
at the followin
Thursday 29 to Fri
Prince Rainier's
Friday 21 to
Aston Mar
Auction
s Aston Martin
May
ams Fine Watch
ednesday 9 June
e Watch Auction,
y 1 to Friday 2 July
wood Festival of Speed
ction, England
esday 7 to Friday 10 Sep
hams Fine Watch Au
ay 17 to Sunday
ember
dwood Revival
ion, England
Enquiries
Paul Maud
+44 (0)
paul.
@bon
.com
aslinge
nams.com
502 200
@bonhams.com
t gold Carrera autom
£7,000 - 9,000
hy by Clemens Kois.
nhams
101 New Bond Street
London W1S 1SR
www.bonhams.com/watches
SKIPPER
HEUER
AUTOMATIC
CHRONOGRAPH
SWISS
SINCE 1860
150
YEARS
LOS ANGELES
MIAMI
NEW YORK
LONDON
PARIS
TOKYO
SHANGHAI
BEIJING
DELHI
MOSCOW
BASEL
MONACO
MILAN
ODY
TAG H

SKIPPER

Die Marke Heuer hatte seit den 1950er Jahren Nautische Chronographen in ihrem Portfolio. Zusätzlich zum Motor- und Wintersport stellt der Rennboot- sowie Wassersport ein wichtiges Segment für einen Hersteller von Sport-Chronographen dar.
Der erste Mann an Bord, der auch die Geschicke auf dem Boot steuert, ist der Skipper – daher stammt auch der Name „HEUER SKIPPER".
Wichtige Faktoren für einen erfolgreichen Regattastart sind das Timing und die Position gegenüber den Konkurrenten beim Durchfahren der Startlinie.
Dazu benötigt man eine sekundengenaue Zeitmessung, um in verbleibenden Minuten die Dauer der Manöver genauso zu berechnen, dass man die Startlinie in voller Fahrt bei idealen Bedingungen passieren kann. Genau auf diese Manöver ist die Heuer Skipper gut vorbereitet. Charakteristisch dafür war das Minuten-Hilfszifferblatt mit der 15-Minuten-Unterteilung und den rot, weiß, blauen Skalierungen. Eine abnehmende Minuteneinteilung erleichtert dabei die Ablesbarkeit vor dem Start.
Alle Skipper Chronographen, bis auf das erste (mit Carrera-Gehäuse, siehe Seite 17/18), sind verwandt mit den Autavia-Modellen und haben die gleichen Gehäuseformen. Die Zifferblätter waren blau oder schwarz mit farblich abgestimmten Minuten- und Stunden-Lünetten.
Die ersten Modelle hatten manuelle Valjoux-Werke 7730 (später 7734 mit Datum), entstanden aus dem eingestellten Modell Venus. 188.Ab 1972 erschien eine Automatikversion, angetrieben durch das neue Kaliber-15-Uhrwerk. Die Automatik Skipper blieb bis Mitte der 1980er im Portfolio. TAG Heuer hat die Tradition der Skipper-Modelle sehr erfolgreich fortgesetzt.

The Heuer Company had nautical stopwatches in its product portfolio from the 1950s. In addition to motorsport and winter sports, sailing competitions and water sports became an important market for a producer of sport chronographs.
The key person on board a boat is the skipper, holding the fate of the vessel in his hands: hence the name "HEUER SKIPPER."
Important success factors during the start of a regatta are the timing and the position in relation to competitors when crossing the starting line. For this task, accurate timekeeping is required so that the boat can cross the starting line at full tilt speed under ideal circumstances. The Heuer Skipper provides this capability. Characteristic of the Skipper were the red, white and blue 15-minute divisions on the sub dial. A minute countdown function adds clarity prior to the start of the race.
All Skipper models except for the very first (in a Carrera case, see page 17/18) are related to the Autavia models, sharing the same case design. The dials were blue or black with color-matched minute and hour bezels. The first models were powered by manual Valjoux 7730 movements (later 7734 with date), developed from the discontinued Venus 188.
From 1972 a self-winding version appeared, driven by a new Calibre 15 automatic movement. The Skipper model series remained in production until the mid-1980s.
TAG Heuer has continued the Skipper Chronographs heritage with enormous success.

SKIPPER

ref. 7764 MH

Modell Skipper (Heuer Katalog 1970) besteht aus dem gleichen Gehäuse mit Druckboden und in beide Richtungen drehbarer Stunden- und Minuten-Lünette wie die Autavia (Referenz 7763). Das mattschwarze Zifferblatt hat zwei Hilfszifferblätter bei 3 bzw. 9 Uhr, jedes von ihnen in Größe und Stil verschieden. „SKIPPER" ist über dem Heuer-Logo und unter 12 Uhr weiß gedruckt.
Das ins Auge springende Kennzeichen ist das große 15-Minuten-Hilfszifferblatt, welches in drei gleich große rote, weiße und blaue Segmente unterteilt ist und einen schwarzen mit Leuchtmasse gefüllten Zeiger hat. Das permanente schwarze Sekunden-Hilfszifferblatt steht bei 9 Uhr. Die zwei Hilfszifferblätter sind versenkt und haben konzentrische Rillen. Die polierten Stahlzeiger haben schwarze sowie aus Leuchtmasse bestehende Einlagen. Der rot bemalte Chronographen-Sekundenzeiger reicht bis zur rot und weiß gedruckten 1/5-Sekunden-Skala. Die metallenen Stundenindizes haben eine gebürstete Oberfläche mit dünnen weißen Einlagen, sind abgeschrägt und haben Leuchtpunkte, die an der äußeren Neigung zur Sekunden-Skala angebracht sind. „SWISS" und „T" für Tritium stehen über dem 6-Uhr-Index. Die Krone ist mit „HEUER" signiert, der Rückdeckel ist voll poliert. Das Uhrwerk ist ein Valjoux Kaliber 7730.

The Skipper model (Heuer brochure of 1970) shares the Autavia reference 7763 case with snap back and bidirectional hour and minute bezel. The matt black dial has two subdials at 3 and 9 o'clock, each of which is sized and styled differently.
"SKIPPER" is printed in white above the Heuer logo below 12 o'clock. The eye-catching feature is the large 15-minute recording subdial, which is divided into three equal-sized red, white, and blue wedges and has a black hand filled with luminous material. The permanent black second subdial is at 9 o'clock. The two subdials are recessed in the dial and have concentric circular ridges. The polished steel hands have black inserts and luminous inlays. The red-painted chronograph second hand reaches the red- and white-printed 1/5-second track at the edge of the dial. Metal hour markers have a brushed surface with thin white inserts. Markers are beveled, with luminous dots applied on the outer bevel next to the second track. "SWISS" and "T" for Tritium are printed above the 6 o'clock index. The crown is fully signed "HEUER." The case back is fully polished. Movement is a Valjoux 7730.

SKIPPER
HEUER
T
SWISS

HEUER
SKIPPER
SWISS

SKIPPER

ref. 73464

Die neue Heuer Skipper (Katalog 1972) besteht aus dem gleichen weiterentwickelten, verschraubten Gehäuse wie das der Autavia, welches die Nachfolge des Kompressor-Gehäuses (Referenz 7764) antritt. Das neue Gehäuse zeigt eine auffällige blaue, in beide Richtungen drehbare, Minuten- und Stunden-Lünette.
Das Zifferblatt in leuchtendem Blau hat zwei Hilfszifferblätter bei 3 und 9 Uhr, die sich in Größe und Gestaltung unterscheiden. Interessant ist die Repositionierung von „SKIPPER" über „SWISS" sowie dem Datumsfenster bei 6 Uhr, während das Heuer-Logo an der gewohnten Position unterhalb von 12 Uhr verbleibt.
Der Blickfang ist ein großes 15-Minuten-Hilfszifferblatt, das in drei gleich große rote, weiße und blaue Segmente unterteilt und dessen schwarzer Zeiger mit Leuchtmasse gefüllt ist. Das Hilfszifferblatt für die permanente Sekunde befindet sich bei 9 Uhr. Beide Hilfszifferblätter sind in das Zifferblatt eingelassen. Die orangefarbenen Zeiger sind mit Leuchtmasse gefüllt. Der orangefarbene Chronographen-Sekundenzeiger reicht bis zur orange und weiß gedruckten 1/5-Sekunden-Skala am Rand des Zifferblatts. Die Stundenindizes aus Metall mit weißen Einsätzen haben Leuchtbalken am äußeren Ende. Die Krone ist mit „HEUER" signiert. Die Drücker sind geriffelt, der äußere Gehäuseboden zeigt einen Längsschliff. Das Uhrwerk ist ein modernisiertes Valjoux 7734 mit der Signatur „7734 R" unter der Unruh sowie „HEUER LEONIDAS SA" auf der Brücke.

The new Heuer Skipper 1972 (Heuer catalog) shares an advanced Autavia screw-back case design succeeding the compressor case, reference 7764. The new design has a striking blue bidirectional minute and hour bezel.
The luminous blue dial has two subdials at 3 and 9 o'clock, each of which is sized and styled differently. Interestingly, the model name "SKIPPER" was relocated above "SWISS" and the date window at 6 o'clock while the Heuer Logo remained in its normal position below 12 o'clock. The eye-catching feature is the large 15-minute recording subdial, which is divided into three equal-sized red, white, and blue triangles and has a black hand filled with luminous material. The permanent second subdial is at 9 o'clock. Both subdials are recessed in the dial. The orange hands have large luminous inlays. The orange-painted chronograph second hand reaches an orange- and white-printed 1/5-second track at the edge of the dial. Metal hour markers with white inserts have luminous bars at the outer end. The crown is fully signed "HEUER." Pushers have a fluted design and the outer case back shows longitudinal grinding. The movement is a modernized Valjoux 7734 signed "7734 R" under the balance and "HEUER LEONIDAS SA" on the bridge.

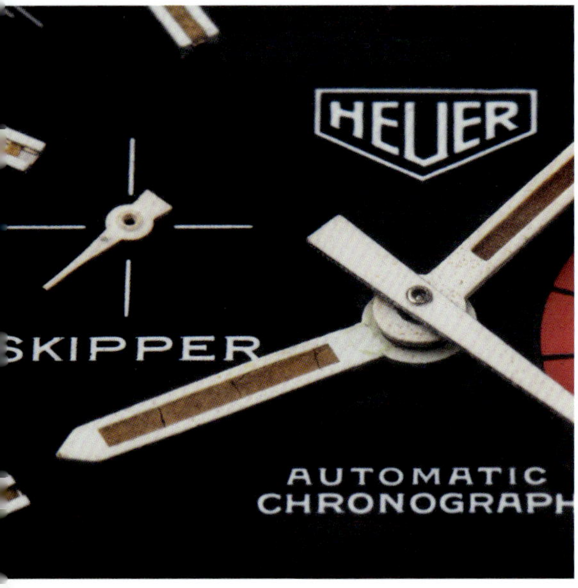

SKIPPER

ref. 15640 „Yachting"

Das letzte Automatik-Modell Skipper (im „Time for Action"-Katalog 1984 Heuer „Yachting" genannt) im endgültigen Autavia-Gehäusedesign (vergrößert auf 42 mm) mit gebürsteter Oberseite sowie großen polierten Seitenflächen und schwarzer Stunden- und Minuten-Lünette, in beide Richtungen drehbar.
Das schwarze Zifferblatt hat eine kleine permanenten Sekunde bei 10 Uhr, unter der bei 9 Uhr die Modellbezeichnung Skipper zu finden ist. Blickfang ist das große, in drei gleich große rote-, weiße- und blaue Segmente unterteilte, 15-Minuten-Zifferblatt. Darüber rotiert ein mit Leuchtmasse gefüllter schwarzer Zeiger. Unter 12 Uhr befindet sich das „HEUER"-Logo, während über dem Datumsfenster bei 6 Uhr in weißer Schrift „AUTOMATIC CHRONOGRAPH" steht. Die weiß gefärbten Stundenindizes aus Metall sind mit Leuchtmaterial gefüllt. Kleinere Indizes befinden sich bei 2, 4 und 9 Uhr. Die weiß lackierten Stahlzeiger haben große Leuchteinlagen. Der weiße Chronographen-Zeiger erreicht die 1/5-Sekunden-Skala am äußeren Rand des Zifferblatts. Die Krone ist auf der linken Seite des Gehäuses positioniert und mit „HEUER" signiert. Die Chronographen-Drücker sind auf der rechten Seite des Gehäuses integriert. Der Gehäuseboden hat einen Längsschliff.
Die Uhr wird von einem Automatikwerk Kaliber 15 angetrieben und ist auf der Brücke mit „HEUER CAL.15" signiert.

Last automatic Heuer Skipper model (listed as the Heuer "Yachting" in the "Time for Action" catalog of 1984), sharing the final Autavia case design (enlarged to 42mm (1.65 in)) with brushed finish on top, large polished sides, and black bidirectional hour and minute bezel. The dial is black with a small permanent second subdial at 10 o'clock and the "SKIPPER" model name underneath at 9 o'clock. The eye-catching feature is the large 15-minute recording subdial, which is divided into three equal-sized red, white and blue triangles with a black luminous filled hand on top. Under 12 o'clock is the "HEUER" logo, while "AUTOMATIC CHRONOGRAPH" is printed in white in white above the date window at 6 o'clock. The white-painted metal hour markers have luminous inserts. Smaller markers are at 2, 4, and 9 o'clock. The white steel hands have large luminous inserts. A white chronograph hand reaches the 1/5-second track on the outer edge of the dial. The crown is on the left of the case and signed "HEUER." Chronograph pushers are integrated on the right of the case. The case back has longitudinal grinding.
The watch is powered by a Calibre 15 automatic movement and is signed "HEUER CAL.15" on the bridge.

HEUER
SKIPPER
AUTOMATIC
CHRONOGRAPH
SWISS
18

HEUER

SOLUNAR

Instruction manual and international service

Mode d'emploi et service international

Bedienungsanleitung und internationaler Service

Printed in Switzerland

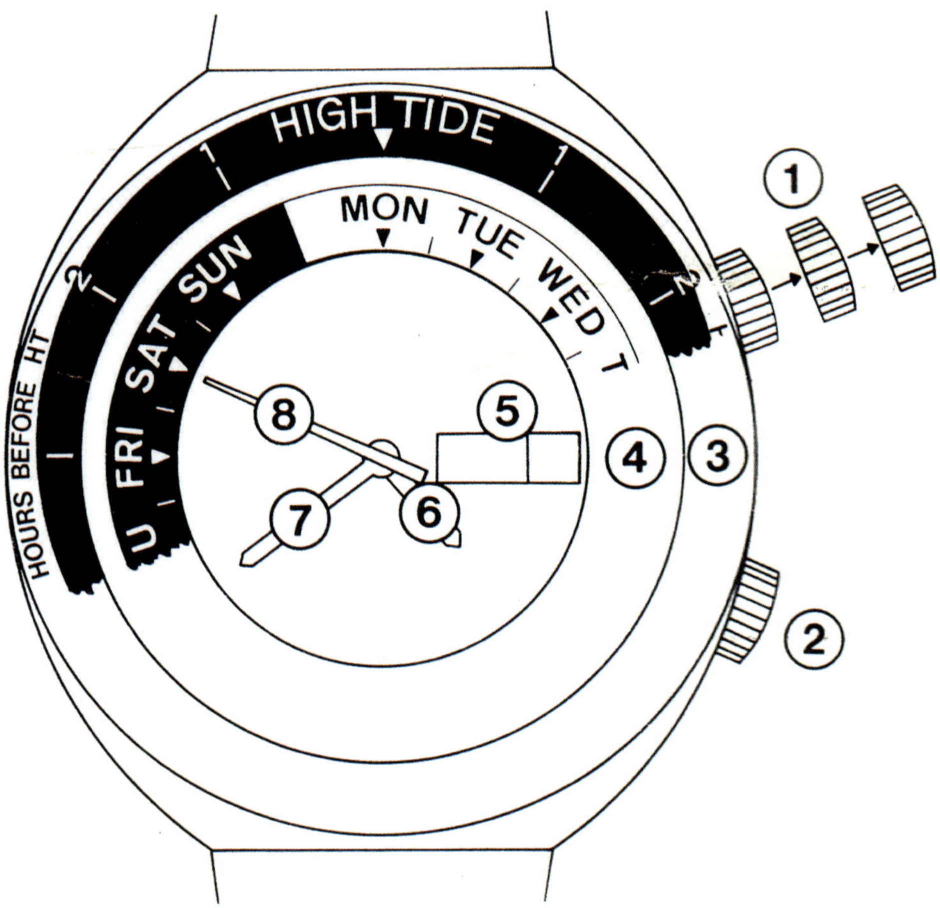

Wie bedient man die HEUER SOLUNAR?

Funktionen

⑦ Richtkrone für Uhrzeiger, Wochentag und Datum
② Richtkrone für inneren Drehring
③ Äusserer Drehring, der den Verlauf der Gezeiten angibt
④ Innerer Datum-Drehring
⑤ Wochentag und Datum
⑥ Stundenzeiger
⑦ Minutenzeiger
⑧ Sekundenzeiger

How to use the HEUER SOLUNAR

Functions

① crown to adjust hands, day and date
② crown to serve inner turning bezel
③ outer turning bezel with tide indications
④ inner turning bezel for date
⑤ day and date
⑥ hour hand
⑦ minute hand
⑧ second hand

SOLUNAR

Sie ist die einzige Uhr in diesem Buch, die keine Chronographen-Funktion besitzt, aber die Messung der Gezeiten ermöglicht. Der Name „SOLUNAR“ setzt sich zusammen aus „SOL“ und „LUNAR“ für Sonne und Mond. Die Heuer Solunar gibt es im Heuer-Portfolio seit den 1950er Jahren. Die hier gezeigte Solunar war die erste Gezeitenuhr mit automatischem Werk. Die Uhr bietet eine zweisprachige Wochentag-Funktion in französischer und englischer Sprache. Die zwei Drehringe müssen alle zwei Wochen neu eingestellt werden. Der Abstand zwischen den Markierungen auf dem inneren Drehring entspricht einer Zeitspanne von 50 Minuten auf dem Zifferblatt. Von Experten wird dieses Intervall als praktischer Mittelwert der täglichen Verschiebung der Gezeiten anerkannt. Auf dem inneren Drehring sind 14 Wochentage angeführt.

Die erste Markierung „MON“ für Montag wird durch Drehen der Richtkrone auf diejenige Stunde des Uhr-Zifferblattes eingestellt, in welcher an diesem Montag „High Tide“, Höchststand der Flut, eintreten wird. Die entsprechende Information steht in Tageszeitungen oder den in Häfen veröffentlichten Gezeiten-Tabellen. Damit ist der innere Drehring eingestellt, und man kann mit dessen Hilfe den Zeitpunkt, in dem „High Tide“ eintreten wird, mühelos auf dem Zifferblatt ablesen, und zwar für jeden Tag innerhalb der kommenden zwei Wochen.

Der äußere Drehring trägt die Gezeiten-Markierungen. Ein Gezeiten-Zyklus dauert ca. 12 Stunden, Ebbe und Flut somit ca. je 6 Stunden. Der Außenring lässt sich so ausrichten, dass er für ein bestimmtes Datum die Flut markiert, wobei die Lage des Innenrings zum Datumsfenster als Referenz dient. Mithilfe des äußeren Drehrings kann man nun den zeitlichen Verlauf von Ebbe und Flut auf dem Zifferblatt ablesen, und zwar für jeden beliebigen Tag innerhalb der kommenden zwei Wochen. Zusätzlich ist die Stundenzahl vor und nach „High Tide“ mittels der auf dem äußeren Drehring aufgetragenen roten Zahlen ablesbar.

Die Heuer Solunar wurde nur in den Jahren 1977 und 1978 in den Katalogen von Heuer beworben und ist als Sammelobjekt extrem selten zu finden.

The Solunar is the only watch in this book that is not a chronograph; it can, however, measure the tides. The name "SOLUNAR" is a portmanteau of "SOL" (sun) and "LUNAR" (moon). The Heuer Solunar has been important in the Heuer portfolio since the 1950s. The Solunar depicted here was the first tidal watch with an automatic movement. The watch provides a bilingual weekday function in French and English. The two bezels must be adjusted every two weeks. The distance between the markings on the inner bezel corresponds to a time span of 50 minutes on the dial. This interval is considered a practical average by experts for the daily change in the timing of the tide. The inner bezel displays 14 days.

The first mark of "MON" for Monday is set by turning the crown to the appropriate hour of the dial on which "high tide" is due to fall on the given Monday. Accurate information as to the timing of the tides should be taken from newspapers or published tide tables found in harbors. The inner bezel is set with this simple operation of the crown and provides a simple reference of the time at which high-tide will begin for each and every day of the week for two weeks' time.

The outer bezel bears tide markers. A tide-cycle lasts approximately 12 hours, with each ebb and flow of the tide approximately 6 hours. The outer bezel can be set to mark "high tide" on a specific date by moving the "HIGH TIDE" mark to the desired date on the inner bezel. The outer bezel enables the ebb and flow of the tide to be followed on the dial for every day within a two-week period. The hours both before and after "high tide" can be read off from the red numbers on the outer ring.

The Heuer Solunar was only advertised in the 1977 and 1978 catalogs and is an extremely rare collector's piece.

SOLUNAR

ref. 279.603

Die Solunar ist eine automatische Gezeitenuhr mit verschraubtem Boden. Sie hat eine massive, schwarze und silberne Lünette mit rotem Aufdruck rund um das Glas, eine polierte Oberfläche und polierte Seiten. Das metallisch-blaue Zifferblatt ist weiß bedruckt, und eine Tages- und Datums-Anzeige steht bei 3 Uhr. Das Zifferblatt ist umgeben von einer angwinkelten inneren Lünette, welche „MON" bis „SUN" anzeigt – eine Woche in Schwarz und eine in Weiß –, in einem silbernen Spannring.
Die zwei Lünetten arbeiten zusammen, um Ebbe und Flut innerhalb einer Periode von zwei Wochen anzuzeigen. Der Modellname „SOLUNAR" steht unter dem Heuer-Logo unter 12 Uhr, „AUTOMATIC" über 6 Uhr. Mit Leuchtmasse gefüllte Stundenindizes sind am Rand des Zifferblattes auf der 60-Sekunden-Skala platziert. Cremefarbene Zeiger haben Leuchtmasse-Einlagen. Zwei Kronen sind auf der rechten Seite des Gehäuses integriert und unsigniert. Der Boden hat einen Längsschliff mit polierten Seiten. Die Uhr wird von einem automatischen ETA Uhrwerk angetrieben und ist signiert.

The Solunar is an automatic tide watch with screw-back case incorporating a rugged black and silver bezel with red print around the crystal and a polished top and sides. The metallic blue dial has white printing and a day/date feature is integrated at 3 o'clock. The dial is surrounded by an angled inner bezel indicating "MON" to "SUN," once in black and once in white, within a silver tension ring.
The two bezels operate together to indicate low and high tide over a period of two weeks. The model name "SOLUNAR" is positioned under the Heuer logo below 12 o'clock. "AUTOMATIC" is printed above 6 o'clock. Luminous hour markers are placed at the edge of the dial on the 60-second track. Cream-colored hands have luminous inserts. Two crowns are integrated on the right of the case and are unsigned. The case back has longitudinal grinding with polished sides. The watch is powered by an automatic ETA movement and signed.

HIGH TIDE
LOW TIDE
HOURS BEFORE HT
HOURS AFTER HT
HEUER
SOLUNAR
AUTOMATIC
JEU 8
SWISS MADE
MON TUE WED THU FRI SAT SUN
SUN SAT FRI THU WED TUE MON

HEUER
HEUER
AUTOMATIC
CHRONOGRAPH
SWISS
18

VERONA

Verona ist eine sehr schöne Stadt im Norden Italiens, in der Region Venetien gelegen. Eingebettet zwischen dem Gardasee im Norden und Venedig im Osten ist die Stadt umgeben von jahrhundertealten Weingütern. Sie hat um die 260.000 Einwohner und ist das wirtschaftliche Zentrum der Region.

Weltberühmt wurde Verona durch die tragische Geschichte des Liebespaares Romeo und Julia aus der Feder von William Shakespeare. Die Opernfestspiele finden jedes Jahr in der Arena statt, einem römischen Amphitheater, das 50 n. Chr. erbaut wurde und 22.000 Besuchern Platz bietet. Das kulturelle Leben und elegante Image der Stadt hatte Jack Heuer bewogen, die neue Modellreihe mit insgesamt 12 Referenzen Verona zu nennen. Dabei wurden 8 Referenzen durch Quarzwerke angetrieben.

Zur Freude der Liebhaber mechanischer Chronographen wurde 1978 die Heuer Verona mit einem Automatik-Werk Kaliber 12 in das Programm aufgenommen.

Das gebürstete Gehäuse wirkte durch klassische Schlichtheit und erinnerte an die schöne Linienführung der frühen Carrera Handaufzugs-Modelle. Die Lünette war entweder aus poliertem Stahl oder dekorativ vergoldet. Verona-spezifisch waren das elegante Zeigerspiel und das klassische Zifferblatt in schwarz-glänzend oder silber, mit einem Armierungsring in gleicher Farbe mit aufgedruckter Tachymeter-Skala.

Die Referenz- und Seriennummern, normalerweise zwischen den Bandanstößen bei 6 und 12 Uhr zu finden, wurden bei den Verona-Modellen auf den Boden gestanzt.

Die Verona blieb zwei Jahre in den Katalogen von Heuer und ist in der Stahlversion ein Höhepunkt an schlichter Eleganz. In der Stahl-Gold-Version mit Goldlünette, vergoldeter Krone und vergoldeten Chronographen-Drückern traf sie besonders den Zeitgeist der späten 1970er Jahre.

Verona is an outstandingly beautiful city in Northern Italy in the Veneto region. Situated between Lake Garda to the north and Venice to the east, the town is surrounded by centuries-old vineyards. Verona has 260,000 inhabitants and is the economic hub of the region.

The tragic tale of Romeo and Juliet, written by William Shakespeare, made Verona world-famous. The opera festival takes place every year in the Arena, a Roman amphitheatre that was built in 50 A.D. and holds 22,000 spectators. The chic cultural image of the city influenced Jack Heuer to dub this new series, with its 12 models, "Verona." Eight of these models were quartz.

Much to the delight of mechanical chronograph aficionados, the Heuer Verona was introduced to the portfolio with a Calibre 12 automatic movement in 1978.

The brushed case was classically sleek and reminiscent of the early Carrera manual-wind models. The bezel was either polished steel or decoratively gold-plated. Specific to the Verona model were the elegantly designed hands and a classic dial, with glossy black or silver face with a matching tension ring on which a tachymeter scale was printed.

The reference and serial number, normally found between the lugs at 6 and 12 o'clock, were placed on the back of the case.

The Verona remained in the Heuer catalogue for two years and the steel version is considered a paragon of sleek elegance. The gold and steel version with gold-plated bezel, crown, and pushers reflected the mood of the late 1970s.

VERONA

ref. 110.215 N

Stahlgehäuse mit verschraubtem Boden, Strichschliff auf der Gehäuseoberseite und seitlich und einer polierten abgerundeten goldenen Lünette. Das Zifferblatt ist glänzend schwarz mit schwarzen 30-Minuten- und 12-Stunden-Hilfszifferblättern, umgeben von einem angewinkelten schwarzen Tachymeter-Spannring. Unter 12 Uhr steht „VERONA" über dem Heuer-Logo, „AUTOMATIC CHRONOGRAPH" über dem gold-gerahmten Datumsfenster bei 6 Uhr.
Die goldenen länglichen Stundenindizes haben eine weiße Auflage, zwei erhöhte Balken und Leuchtpunkte am äußeren Ende. Ein paralleler Index steht bei 12 Uhr, keine Indizes bei 3, 6 und 9 Uhr. Die zarten goldenen Zeiger mit Leuchtmaterial haben glänzend weiße Ränder. Der weiße große Chronographen-Sekundenzeiger berührt die 1/5-Sekunden-Skala am Rand des Zifferblattes. Die goldene Krone ist links und mit „HEUER" signiert. Die goldenen Chronographen-Drücker sind rechts integriert, aber auf der Rückseite vollkommen sichtbar. Der Boden ist mit der Referenz- und Seriennummer signiert und hat einen Längsschliff.
Das Uhrwerk ist ein Kaliber 12 und ist mit „CAL. 12" auf der Brücke signiert.

The case is screw-back with brushed finish on top and sides and rounded polished gold bezel. The dial is glossy black with 30-minute and 12-hour recording subdials surrounded by an angled black tachymeter tension ring. Underneath 12 o'clock, "VERONA" stands over the Heuer logo and "AUTOMATIC CHRONOGRAPH" is above the gold-framed date window at 6 o'clock. Elongated golden hour markers have a white coating on two raised bars and luminous dots at the outer end. A parallel marker is at 12 o'clock and markers have been omitted at 3, 6, and 9 o'clock. Fine gold-plated hands with luminous material have a gleaming white edging at the outer sides. The white chronograph hand reaches the 1/5-second track on the outer edge of the dial. The golden crown is integrated on the left of the case and signed "HEUER." Golden chronograph pushers are integrated on the right of the case, fully visible from the back. The case back is signed with a reference and serial number and has longitudinal grinding.
The movement is a Calibre 12, and is signed "CAL.12" on the bridge.

VERONA
HEUER
AUTOMATIC
CHRONOGRAPH
SWISS
TACHY
18

TACHY
VERONA
HEUER
AUTOMATIC
CHRONOGRAPH
SWISS

VERONA

ref. 110.215 G

Stahlgehäuse mit verschraubtem Boden, Strichschliff auf der Gehäuseoberseite und seitlich und einer polierten abgerundeten goldenen Lünette. Das glänzend silberne Zifferblatt hat einen leichten vertikalen Schliff und 30-Minuten- sowie 12-Stunden-Hilfszifferblätter, umgeben von einem angewinkelten silbernen Tachymeter-Spannring. Unter 12 Uhr steht „VERONA" über dem Heuer-Logo, „AUTOMATIC CHRONOGRAPH" über dem stahlgerahmten Datumsfenster bei 6 Uhr.

Die länglichen Stundenindizes haben eine schwarze Auflage aus zwei erhöhten Balken und Leuchtpunkte am äußeren Ende. Ein paralleler Index steht bei 12 Uhr, keine Indizes bei 3, 6, und 9 Uhr. Die zarten goldenen Zeiger mit Leuchtmasse haben schwarze Ränder. Der schwarze Chronographen-Sekundenzeiger berührt die 1/5-Sekunden-Skala. Die goldene Krone ist auf der linken Seite des Gehäuses integriert und mit „HEUER" signiert. Die goldenen Chronographen-Drücker sind rechts integriert, aber auf der Rückseite vollkommen sichtbar. Der Boden ist mit der Referenz- und Seriennummer signiert und hat einen Längsschliff. Das Uhrwerk ist ein Kaliber 12 und ist mit „CAL. 12" auf der Brücke signiert.

The case is screw-back with brushed finish on top as well as sides and rounded polished gold bezel. The gleaming silver dial has light vertical grinding with 30-minute and 12-hour recording subdials surrounded by an angled silver tachymeter tension ring. Underneath 12 o'clock, "VERONA" is displayed over the Heuer logo and "AUTOMATIC CHRONOGRAPH" is above the gold-framed date window at 6 o'clock. Elongated golden hour markers have a black coating on two raised bars and luminous dots at the outer end. A parallel marker is at 12 o'clock and markers have been omitted at 3, 6, and 9 o'clock. The fine gold-plated hands with luminous inserts have black edging. The black chronograph hand reaches the 1/5-second track on the outer edge of the dial. The golden crown is integrated on the left of the case and signed "HEUER." Golden chronograph pushers are integrated on the right of the case, fully visible from the back. The case back is signed with a reference and serial number and has longitudinal grinding. The movement is a Caliber 12, and is signed "CAL.12" on the bridge.

VERONA

ref. 110.213 N

Stahlgehäuse mit verschraubtem Boden, Strichschliff auf der Gehäuseoberseite und seitlich und einer polierten abgerundeten Lünette. Das Zifferblatt ist glänzend schwarz mit schwarzen 30-Minuten- und 12-Stunden-Hilfszifferblättern, umgeben von einem angewinkelten schwarzen Tachymeter-Spannring. Unter 12 Uhr steht „VERONA“ über dem Heuer Logo, „AUTOMATIC CHRONOGRAPH“ über dem stahlgerahmten Datumsfenster bei 6 Uhr. Die länglichen Stundenindizes haben eine silberne Auflage, zwei erhöhte Balken und Leuchtpunkte am äußeren Ende. Ein paralleler Index steht bei 12 Uhr, keine Indizes bei 3, 6, und 9 Uhr.
Die zarten silbernen Zeiger mit Leuchtmaterial haben glänzend weiße Ränder. Der weiße große Chronographen-Sekundenzeiger berührt die 1/5-Sekunden-Skala. Die Krone ist links und mit „HEUER“ signiert. Die Chronographen-Drücker sind rechts integriert, aber auf der Rückseite vollkommen sichtbar. Der Boden ist mit der Referenz- und Seriennummer signiert und hat einen Längsschliff.
Das Uhrwerk ist ein Kaliber 12 und ist mit „CAL. 12“ auf der Brücke signiert.

The case is screw-back with brushed finish on top and rounded polished bezel. The dial is glossy black with 30-minute and 12-hour recording subdials surrounded by an angled black tachymeter tension ring. Below 12 o'clock, "VERONA" is shown over the Heuer logo and "AUTOMATIC CHRONOGRAPH" is above the steel-framed date window at 6 o'clock.
The elongated hour markers have a silver coating on two raised bars and luminous dots at the outer end. A parallel marker is at 12 o'clock, and markers have been omitted at 3, 6, and 9 o'clock. Fine silver hands with luminous inserts have glossy white edging. The white chronograph hand reaches the 1/5-second track on the outer edge of the dial. The gold-plated crown is integrated on the left of the case and signed "HEUER." Steel chronograph pushers are integrated on the right of the case, fully visible from the back. The case back is signed with a reference and serial number and has longitudinal grinding.
The movement is a Caliber 12, and is signed "CAL.12" on the bridge.

HEUER
TACHY
VERONA
HEUER
AUTOMATIC
CHRONOGRAPH
SWISS

HEUER
JACKY ICKX
EASY-RIDER
CHRONOGRAPH
SWISS MADE
TACHY

JACKY ICKX

Um neue Kunden zu gewinnen, entschied Jack Heuer im Jahre 1972, sich auf die jüngere Klientel zu konzentrieren, die an Sport interessiert und daher für das Tragen von Chronographen empfänglich ist. Dazu erneuerte er den Sponsorenvertrag mit dem Ferrari-Rennteam. Das Heuer-Logo wurde deshalb in markanter Weise auf den Ferrari-Rennautos positioniert, die in der Formel 1 und in der Sportwagenserie konkurrierten. Die Fahrer Jacky Ickx, Clay Regazzoni und Ronnie Peterson wurden ausgewählt, das Heuer-Chronograph-Zeichen auf ihren Rennanzügen zu tragen.
Jacky Ickx, der berühmte junge belgische Fahrer, wurde als Repräsentant der neuen Chonographen-Linie ausgewählt, die seinen Namen und seine Unterschrift am Zifferblatt zeigte. Die Notwendigkeit von Zeitmessern im Motorsport war ideal, um das neue Produktmodell während der Rennsaison weltweit anzupreisen – eine großartige Gelegenheit für positive Imagetransfers. Ein einzigartiger Miniaturnachbau des Helms von Jacky Ickx diente als Etui, welches vom Verkaufsgesichtspunkt höchst attraktiv war. Der Chronograph wurde der Öffentlichkeit auf der Basler Uhrenmesse präsentiert. Heuer lancierte mit dem „Jacky Ickx Easy Rider"-Modell eine preisgünstige Alternative für junge Interessenten, folglich eine ideale Einstiegsstrategie für potenzielle zukünftige Kunden.
Das verchromte Gehäuse mit Tachymeter-Skala enthält ein Stiftankerwerk mit Datumsanzeige und 15-Minuten-Totalisator. Vier verschiedene Farbkombinationen des Ziffernblattes waren erhältlich: schwarz, blau, rot und weiß.
Mit der Leonidas Easy Rider lancierte Heuer eine billigere Variante mit Gehäusen aus Glasfaserkunststoff (GFK). Design und Größe waren ident mit der Jacky Ickx. Beide Uhren trieb ein Chronographen-Werk Ebauche Bettlach Kaliber 8420 an, jedoch fehlte die Datumsfunktion bei der Leonidas Easy Rider. Anschließend folgte die Heuer Temporada mit Valjoux 7733-Werk im GFK-Gehäuse und Metall-Lünette. Die seltenste Variante der Temporada trägt eine auffällige Zielflagge auf dem Zifferblatt.

To attract new customers, in 1972 Jack Heuer decided to focus on younger consumers with an interest in sports, and therefore receptive to the use of chronographs. Concurrently, he renewed the sponsorship contract with the Ferrari racing team. Hence, the Heuer logo was positioned prominently on Ferrari racing cars competing in Formula One and sports car series. The drivers Jacky Ickx, Clay Regazzoni and Ronnie Peterson were chosen to display the Heuer Chronograph badge on their racing suits.
Jacky Ickx, the famous young Belgian driver, was chosen as ambassador for a new chronograph line bearing his name and signature on the dial. The close link to the need for timekeeping devices in motorsport was ideal to promote the new product model line around the world during the race season, offering a great opportunity for positive image transfers. A unique miniature replica of Jacky Ickx's racing helmet formed the case and was an extremely attractive sales draw.
The chronograph was presented to the public at the Basel watch fair. Heuer's "Jacky Ickx Easy Rider" model was a low-cost alternative for young clients to experience the brand—an ideal entry strategy for potential future customers.
The chromed metal case with tachymeter division houses a pin-pallet movement featuring date and a 15-minute subdial.
Four different color combinations for the striking dial were available: black, blue, red, and white.
Heuer's Leonidas Easy Rider was a cheaper version with a fiberglass plastic (GRP) case. The design and size were identical to the Jacky Ickx. Both watches were powered by an Ebauche Bettlach caliber 8420 chronograph movement, but the Leonidas Easy Rider lacked the date function. This was followed by the Heuer Temporada, with a Valjoux 7733 movement in a GRP case and metal bezel. The rarest version of the Temporada has an eye-catching checkered flag motif on the dial.

Chronograph Jacky Ickx Easy-Rider, signiert von Jacky Ickx am 15.2.1973.

Jacky Ickx Easy-Rider Chronograph signed by Jacky Ickx on the 15th of February 1973.

Jacky Ickx

LEONIDAS
EASY-RIDER
CHRONOGRAPH
SWISS MADE
HOBIE CAT
CHRONOGRAPH
TACHY

HEUER
TEMPORADA
SWISS
TACHYMETRE
BASE 15 PULSATIONS
HEUER Temporada
Ref. 733.809 R Temporada C
graph, fiberglasverstärktes Kunst
gehäuse mit Stahl-Lünette, wass
dicht, rotes Leuchtzifferblatt, Pul
733.806 CH Temporada
nograph, fiberglasverstärktes
stoffgehäuse mit Doublé-
te, wasserdicht, champagne

HEUER
CARRERA
Automatic Chronograph
World wide service
Service mondial

HEUER
HEUER
Garantie
Garantia

1163 MH
HEUER
HEUER
AUTAVIA
HEUER
Servicio mundial
Internationaler Service
AUTOMATIC

HEUER
CAMARO

ACCESSOIRES/ *ACCESSORIES*

HEUER BOXES

Für jeden Sammler ist es etwas Besonderes, wenn er eine alte Uhr mit originaler Box findet. Ist es doch oft ein Hinweis, dass die Uhr und ihre Accessoires geschätzt und gut aufbewahrt wurden. Es gibt auch ungetragene Stücke, die hoch im Kurs stehen und sehr selten sind. Die Box gibt Schutz, sieht hochwertig aus und ist ein wichtiges Requisit, um die Marke zu präsentieren. Sie gibt dem Konsumenten das Gefühl, etwas Elegantes, qualitativ Hochwertiges oder Trendiges gekauft zu haben, weshalb oft ein Markenwert oder Produktattribut kommuniziert wird: zum Beispiel „HEUER SWITZERLAND“ für Hochwertigkeit oder „HEUER AUTOMATIC“ für das Novum Automatik-Chronograph. Die Aktivität von Heuer als Sponsor im Rennsport führte zur Entwicklung von Verpackungen in der Form kleiner Helme. So wurde für die Jacky-Ickx-Chronographen das Helmdesign des Fahrers detailgenau übernommen. Gut erhaltene Exemplare erreichen mittlerweile auf den Sammelbörsen Höchstpreise und sind immer sofort verkauft. Im Zuge der erweiterten Modellpolitik wurde von Heuer seit 1963 eine interessante Vielfalt unterschiedlicher Boxen entwickelt.

Für Carrera- und Autavia-Modelle gab es eine frühe längliche Metall-Box Sie ist außen mit rotem Papier beklebt, das Heuer-Logo in Gold ist mittig platziert, zwei verdeckte Gelenke sind hinten angebracht, ein vergoldeter Metallrand an der Öffnung. Innen ist sie mit rotem Cordsamt ausgelegt und hat zwei Halterungen. Den Deckel bedeckt cremefarbiges Samtpapier mit dem Heuer-Logo in Gold, darunter steht der Schriftzug „SWITZERLAND“.

Die elegante Holzbox hat einen Bezug aus rotem Kunstleder mit schwarzer Maserung. Das Heuer-Logo ist vorn in die Mitte geprägt, umgeben von zwei dünnen goldenen Streifen. Innen sieht man lackiertes Holz. Die Präsentationsspange liegt auf einer mit rotem Samt

All collectors feel a thrill when they find an old watch together with the box it came in. It often indicates that the watch and its accessories were appreciated and looked after by its owner. Some mint condition watches, survive extremely rare and very expensive. The box provides protection, has a premium appearance, and is an important factor in presenting the brand name. It gives the consumer the feeling of purchasing an elegant, high-quality product which is also fashionable. For this reason boxes often communicate the brand name and product attributes: for example, "HEUER SWITZERLAND" is a synonym of good quality, and "HEUER AUTOMATIC" for innovative automatic chronographs. The invention of small helmets as packaging stems from Heuer's sponsorship activities in motorsport. The Jacky Ickx chronographs came in a detailed miniature copy of his helmet. Well-preserved examples have fetched high prices at collectors' markets and are always sold immediately.

As Heuer advanced its portfolio strategy, the company began to develop an interesting variety of boxe, starting in 1963.

On an early oblong metal box for the Carrera and Autavia models, the outside is covered in red paper displaying the Heuer logo in gold in the centre, with two concealed hinges at the back, a gold-plated metal rim at the opening, red corduroy lining, and two mounts. The lid is lined with cream-coloured velvet paper embossed with the Heuer logo in gold and below the word "SWITZERLAND."

The elegant wooden box is covered in red faux leather with black grain. The Heuer logo is embossed centrally on the front, surrounded by two thin gold lines. The interior is of lacquered wood. The display mount is fixed to a red velvet base; the watch is presented flat, which also works for steel bracelets. The box was used for

bezogenen Basis, zeigt die Uhr liegend und erlaubt auch die Präsentation von Uhren mit Stahlbändern. Diese Box wurde sowohl für Handaufzug- als auch für die ersten Automatik-Modelle am Ende der 1960er Jahre verwendet.

Die wohl bekannteste Heuer-Box ist die rote rechteckige mit dem Zielflaggenstreifen. Es existieren zwei Versionen: mit und ohne „AUTOMATIC"-Prägung unter dem Zielstreifen. Eingeführt wurde sie 1969/70 mit den ersten Automatik-Modellen. Es handelt sich um eine hochwertige Kartonbox mit festem Klappmechanismus. Geschlossen prangt das Heuer-Logo vom Vorderteil des Deckels, darunter der Zielstreifen und die Prägung „AUTOMATIC", weiß auf rotem Grund. Geöffnet ist ein schwarzes Heuer-Logo auf weißem Samt zu sehen – in der gleichen Größe wie außen. In der Basis befindet sich ein Kartoneinsatz, der mit rotem Samt überzogen ist und eine verstellbare Samtspange in der Mitte eingeklemmt hat, damit auch eine Uhr mit Stahlband fixiert werden kann. Die früheren Boxen haben eine Polsterung am Deckel und der „AUTOMATIC"-Schriftzug ist feiner ausgeprägt; auch leichte Unterschiede beim Auftrag des Zielstreifens kann man beobachten. Es gibt auch Boxen ohne Zielstreifen. Die abgebildete Box hat noch den Überkarton mit der aufgestempelten Referenz 1163 MH, die zeigt, dass sie eine Autavia 1163 mit Stunden- und Minuten-Lünette „beherbergt". Diese Box wurde speziell für die Autavia-, Carrera- und Monaco-Modelle konzipiert, bei den Handaufzugsmodellen entfiel der „AUTOMATIC"-Schriftzug.

Eine neue Variante folgte mit dem länglichen Autavia-Karton von 1971. Die Oberseite zeigt eine angeschnittene silbern-schwarze Autavia Referenz 1163 MH mit Stahlband von Gay Frères auf rotem Untergrund. Im unteren Drittel steht das Heuer-Logo weiß auf rot. Das

both manual-wind and self-winding models at the end of the 1960s.

The most popular is the rectangular red box showing a band of checkered flags. There are two versions, with and without "AUTOMATIC" embossed below the band. It went into production in 1969/70 for the first automatic models. It was a high-quality fliptop case; when closed the Heuer logo graced the front of the lid. Below the chckered band, the word "AUTOMATIC" is embossed, white on red. When open, the box displays a black Heuer logo, identical in size to the external logo, on white velvet. The base has a cardboard inset covered in red velvet; an adjustable velvet mount is fixed to the center which can also hold a watch with a steel bracelet. The early boxes have a padded lid and more delicate "AUTOMATIC" lettering. Subtle differences in the application of the checkered band can be observed. The box pictured here has an outer casing with the reference 1163 MH, indicating that it "accommodates" an Autavia 1163 with hour and minute bezel. This box was designed especially for the models Autavia, Carrera, and Monaco; for manual-wind models this box was used without the "AUTOMATIC" text.

In 1971, a new variant arrived with the oblong Autavia carton. The upper half depicts a cropped silver-black Autavia, reference 1163 MH, with Gay Frères steel band on a red background. The Heuer logo is shown on the lower third of the box, printed white on red. The lower half is black, and the inlay with pocket is white. The Heuer logo can be found on the short front side, along with a space reserved for the reference and "MADE IN SWITZERLAND" below.

The classic large blue Heuer box was often used in the 1970s. It is elegant, with an embossed Heuer logo on the front. When open, it displays a presentation surface

HEUER
HEUER

HEUER
HEUER
HEUER
HEUER

Unterteil ist schwarz und das Inlay weiß. Am kleinen Rechteck der Vorderseite ist nochmals das Heuer-Logo und Platz für die Referenz ausgespart. Darunter steht „MADE IN SWITZERLAND".

Die klassische große blaue Heuer-Box ist in den 1970er Jahren oft verwendet worden. Sie ist sehr elegant und mit einem schön geprägten Heuer-Logo an der Stirnseite versehen. Geöffnet zeigt sich eine Präsentationsfläche mit Spange, die mit blauem Samt bezogen ist. Das Oberteil ist mit weißem Samt ausgekleidet.

Die rote Klappbox ist kleiner als die vorher beschriebene: Sie ist aus Plastik und mit Lederimitat überzogen. Das Heuer-Logo ist außen in die Mitte des Deckels geprägt. Ein kleiner goldfarbener Punkt zeigt an der Vorderseite des Deckels den Öffnungspunkt an, vergleichbar mit einem Schmuckkästchen. Innen ist roter Cordsamt auf zwei Ebenen ausgelegt und mit einer Aussparung versehen, um das Uhrenband fixieren zu können.

Die wohl „coolste" Uhrenverpackung, die ich kenne, ist der rote Heuer-Helm. Aus Kunststoff gefertigt, ist das edle Stück dem Original eines Vollvisierhelms als Miniatur nachempfunden. Wie das Original hat auch er ein Plastikvisier, das bei Bedarf geschlossen werden kann. Der große ovale Sehschlitz ist mit weißem Klebeband und schwarzem Gummi eingefasst – das Gleiche wiederholt sich zur Basis hin. Das Innenleben besteht aus einem weißen Podium mit einer Kunststoffspange als Uhrenhalter. An der Stirnseite sowie an den Schläfen befinden sich Heuer-Logos. Der Helm kann mühelos von der Basis abgenommen werden, um guten Zugriff zur Uhr zu bekommen. Meist wurden die Rennsport-Modelle wie Carrera, Daytona, Modena, Monza mit Helm ausgeliefert.

with a blue velvet-covered mount and gold rim. The interior of the lid is lined with white velvet.

This red lidded box is smaller than the one described above: it is made of plastic and covered with faux leather. The Heuer logo is engraved in the middle of the outer lid. A small golden dot indicates the opening point of the box, similar to a jewelry box. The interior is lined with red corduroy over two tiers and has a recess to hold the watch band.

The coolest watch presentation box that I know is the red Heuer racing helmet. Made of plastic, this stylish piece is a miniature of the real thing. Like the original, the miniature has a plastic visor that can be opened and closed. The large oval cutout (actually a window into the presentation box) is edged in white tape and black rubber, a pattern repeated to the base. The interior boasts a white podium with a plastic mount for fixing the watch. The front and sides are decorated with the Heuer logo. The helmet can easily be removed from the base to access the watch. The motorsport models, such as the Carrera, Daytona, Modena, and Monza, were supplied in these helmets.

HEUER
CARRERA

AUTAVIA
HEUER
AUTOMATIC
CHRONOGRAPH

HEUER
VERONA

HEUER
SUISSE
SPEZIALIST
FÜR CHRONOGRAPHEN
STOPPUHREN UND
BORDINSTRUMENTE
OFFIZIELLE VERTRETUNG
HEUER
AUTAVIA HEUER
AUTAVIA HEUER AUTOMATIC
CARRERA HEUER AUTOMATIC
CARRERA HEUER
Zeitmesser für Offizielle
Sport-Zeitmessung
Yachting

HEUER
Heuer macht keine gewöhnlichen Uhren.
Der Beweis:
Montreal
utomatische
rmbanduhr
it
ingebauter
topp-
nrichtung.
ef.
10.503
Monte Carlo
Robuste Bord-Stoppuhr für
Auto, Boot und Flugzeug.
Ref. 542.701
Calculator
Automatischer
Chronograph mit
Präzisions-
Rechenschieber.
Ref. 110.633
Can-Am
reisgünstige
leuer-Stoppuhr,
nodernes
tyling.
ef. 804.501
Fibershell
Kurzzeitmesser.
Griffsicheres,
säurebeständiges
und wasser-
geschütztes
Microsplit
Eine vollelektro-
nische Taschen-
Digital-Stoppuhr.
Ref. HL 802
Bezugsquellennachweis: GEBRÜDER JOCHUM, 6901 Bregenz, Postfach 10

GERD-RÜDIGER LANG

Das erste Mal hatte ich im Sommer 2007 die Gelegenheit, Gerd-Rüdiger Lang kennenzulernen.
Ich befand mich aufgrund einer Wettbewerbsklausel in einer Auszeit zwischen zwei Jobs, daher war es der perfekte Zeitpunkt, um ein neues Buchprojekt zu starten: „Heuer Chronographs and Motorsports".
Marcella Prior-Callwey vom Callwey-Verlag holte Gerd-Rüdiger Lang hinzu, um die Uhren zu überprüfen, die in meinem Buch vorgestellt werden sollten, da er jahrzehntelang bei Heuer gearbeitet hatte.
Ein paar Tage später kam ich voller Neugier und Begeisterung am Hauptsitz von Chronoswiss in Karlsfeld am Stadtrand von München an. Das Gebäude war imposant, lichtdurchflutet, minimalistisch-modern mit einem klassischen Touch. Eine Dauerausstellung historisch bedeutender Chronographen fiel mir ins Auge, die alle in speziell angefertigten kleinen Fenstern präsentiert wurden, die über die Treppenhäuser und das gesamte Gebäude verteilt waren.
Gerd-Rüdiger Lang kam aus seinem Büro im ersten Stock herunter. Er war groß, trug einen blauen Blazer, eine rote Hose und Maßschuhe.
„Hallo Herr Haslinger, Sie sind wegen Heuer hier", sagte er zu mir und reichte mir kurz die Hand. Ohne Umschweife betraten wir einen Besprechungsraum in der Nähe der Rezeption.
Ich erkannte schnell, dass Zeit für Gerd-Rüdiger Lang ein Privileg war, weshalb wir meine ausführliche Präsentation von 81 Chronographen zügig durchgingen. Er teilte sein beeindruckendes Wissen ungefiltert mit mir, wofür ich sehr dankbar war.
Nach Abschluss des Heuer-Tutorials sagte er: „Ich habe gehört, dass Sie bei Jaguar in England gearbeitet haben", und stellte sogleich mein Wissen über den Jaguar E-Type auf die Probe, was glücklicherweise gut lief. So bekam ich einen Eindruck von einem Keller voller Jaguars und eine Einladung, an einem anderen Tag wiederzukommen.
Einerseits war Gerd-Rüdiger Lang ein Gentleman von nobler Zurückhaltung, andererseits konnte er mit seinem trockenen Humor unglaublich lustig sein.

The first time I had the opportunity to meet Gerd-Rüdiger Lang was in the summer of 2007.
I was on a timeout between two jobs due to a competitor clause, hence, perfect timing for kicking off a new book project: Heuer Chronographs and Motorsports.
Marcella Prior-Callwey from Callwey publishing house brought in Gerd-Rüdiger Lang to verify the watches to be featured in my book, as he was a veteran member of Heuer for decades.
A few days later, full of curiosity and enthusiasm, I arrived at the Chronoswiss headquarter in Karlsfeld on the outskirts of Munich. The building was imposing, flooded with light, in a minimalistic modern style with a classic twist. A permanent display of historically important chronographs caught my eye, all displayed in specially crafted little windows dispersed around staircases and throughout the building.
Gerd-Rüdiger Lang came down from his first-floor office. Tall, dressed in a blue blazer, red pants, and handmade shoes.
"Hello, Mr. Haslinger. You're here because of Heuer," he said with a quick handshake and immediately led me to a meeting room close to reception.
I soon realized that for Gerd-Rüdiger Lang, time was a privilege; hence, we rushed through my detailed presentation of 81 chronographs without delay. He shared his impressive knowledge directly, and I felt grateful for receiving such excellent feedback.
After finishing his Heuer "tutorial," he said: "I hear you worked at Jaguar in England?" and promptly tested my knowledge about the Jaguar E-type, which fortunately went well. As a result, I got a glimpse of a cellar full of Jaguars as well as an invitation to come back another day.
Gerd Rüdiger Lang was a gentleman with a reserved, aristocratic nature; but he could also be hysterically funny with his ultra-dry sense of humor.

Joseph Siffert
Place de la Gare
1700 Fribourg, Suisse

Tél.: (037) 27710 – 27740
29099
Câble: Siffertauto
Télex: 36265 sifer ch
Banque: Union de Banques Suisses – Fribourg
C.C.P.: 17-4864

Herrn LANG
IWC Uhren GmbH Deutschland
Abt. Heuer-Leonidas
Rossmarkt 14
6 FRANKFURT/ Main

Deutschland

Eingegangen
1700 Fribourg, 10. September 1970
16. SEP. 1970

Sehr geehrter Herr Lang,

besten Dank für Ihren freundlichen Brief vom 19. August 1970. Ich möchte es hiermit nicht unterlassen Ihnen noch einmal für den freundlichen Empfang in Frankfurt zu danken wie auch für die mir in Ihrem oben erwähnten Brief angebotene Unterstüzung, betreffend Ausstellungen.

Was nun die in Ihrem Brief erwähnte Ausstellungen betrifft, teile ich Ihnen mit, dass ich diese Idee bis auf nächstes Jahr verschoben habe. Ich werde Ihnen selbstverständlich zur gegebener Zeit Bericht geben.

Das Bild von Steve McQueen, das ich Ihnen in Frankfurt zeigte, gab ich Herrn J. Heuer, der Ihnen dann selber einen Abzug dieses Bildes übergeben wird.

Die benötigten rot-blau karierten Mützen, kann ich Ihnen leider nicht zustellen, denn meine Mütze erhielt ich als Geschenk. Sollte ich aber nächsthin selber eine "Siffert-Mütze" herausgeben, werden Sie selbstverständlich sofort eine erhalten..

Mit freundlichen Grüssen

Eingegangen

Jo Siffert

Beilagen: 2 signierte Bilder

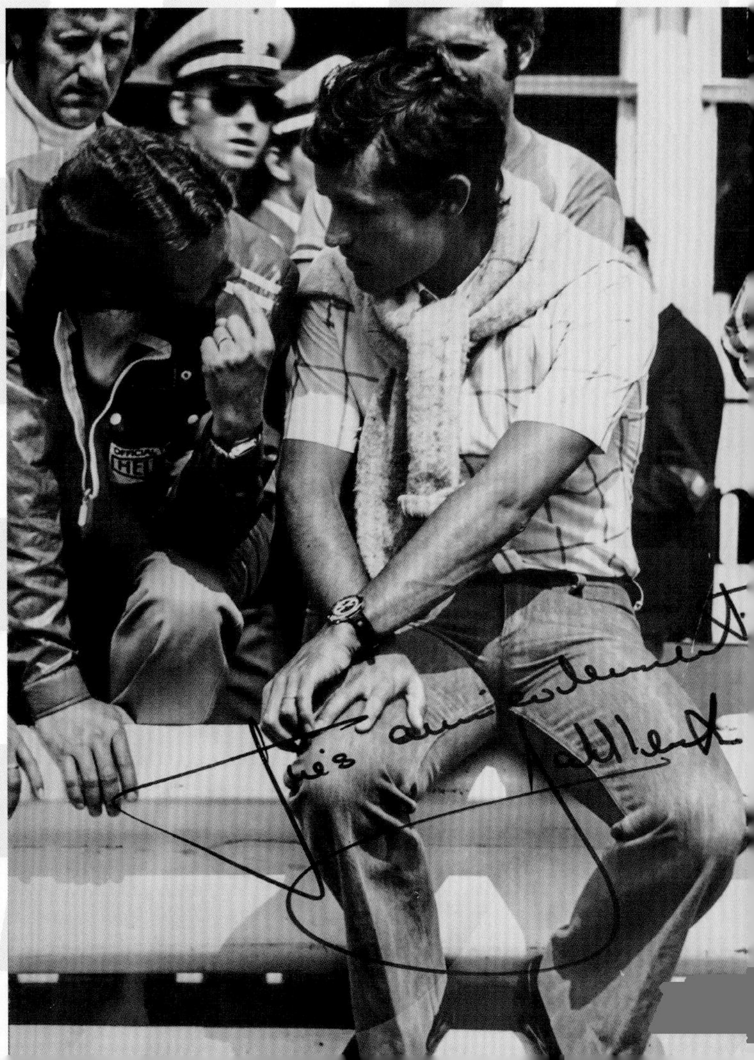

Aber wer war Gerd-Rüdiger Lang? Er wurde 1943 in der deutschen Industriestadt Braunschweig geboren. Als Jugendlicher hatte er kein großes Interesse am Studieren, sondern zog es vor, das Handwerk des Uhrmachers zu erlernen.

Nach einer kurzen Zeit bei der Uhrenfirma Hamilton begann der junge Geselle Mitte 1964 bei Heuer in Biel zu arbeiten. Zeitgenössische Fotos zeigen Herrn Lang bei der Arbeit an der Montagestation für Stoppuhren, dem Hauptgeschäftsbereich von Heuer.

1968 kehrte Lang gerne nach Deutschland zurück und übernahm neue Aufgaben im gemeinsamen Servicezentrum von Heuer und IWC in Frankfurt.

Nur ein Jahr später brachte Heuer in Zusammenarbeit mit Breitling, Buren und Hamilton den ersten Automatik-Chronographen auf den Markt.

Gerd-Rüdiger Lang wurde von Jack Heuer gebeten, eine Kiste voller neuer Uhren nach Frankreich zu bringen, die in dem Film „Le Mans" mit Hollywood-Legende Steve McQueen zu sehen sein sollten (Entwickelte sich zu einer fortwährenden Werbekampagne für die Heuer Monaco). Am nächsten Morgen machte er sich in seinem MG auf den Weg nach Le Mans und schnupperte echte Rennatmosphäre, die so süchtig machen kann. Gerd-Rüdiger Lang stürzte sich kopfüber in die Aufgabe und entdeckte sein Talent, in den kommenden Jahren Zeitnehmer für die Olympischen Spiele und den Motorsport zu werden. Die Split-Seconds Chronographen Nr. 63 und Nr. 67 von Heuer waren seine persönlichen Lieblings-Zeitnehmeruhren.

Später im Jahr 1974 zog das Heuer-Servicecenter nach München um. Gegen Ende der 1970er Jahre wurde die gesamte Uhrenindustrie von der Quarzkrise hart getroffen, daher beschloss Heuer 1980, seine deutsche Niederlassung zu schließen. Im selben Jahr beendete Gerd-Rüdiger Lang seine Ausbildung zum Uhrmachermeister und erhielt das Angebot, den Service für Heuer-Uhren in Deutschland zu übernehmen. Daraufhin gründete er seine eigene Werkstatt in München und machte sich selbstständig.

Gerd-Rüdiger Lang wurde von Jack Heuer als offizieller Zeitnehmer zu den Olympischen Spielen in Moskau eingeladen. Dafür erhielt er eine orange personalisierte Heuer Senator GMT, versehen mit dem Logo von Olympia Moskau 1980, als Geschenk.

Im Jahr 1983 gründete er Chronoswiss und ließ die Marke 1984 eintragen. Gerd-Rüdiger Lang hatte ein großes Talent dafür, die Zukunft vorauszusehen, und glaubte an die Renaissance mechanischer Uhrwerke, als fast niemand sonst daran glaubte.

Er ging davon aus, dass eine Quarzuhr nach wenigen Jahren im Müll landet, während eine mechanische Uhr nach 100 Jahren in einem Museum zu sehen sein wird!

Zweifellos kann Gerd-Rüdiger Lang als einer der ersten und erfolgreichsten unabhängigen Uhrmacher angesehen werden. Er war mir bis zu seinem Tod am 2. März 2023 in Starnberg, Deutschland, ein großartiger Mentor.

Er wird uns in guter Erinnerung bleiben.

But who was Gerd-Rüdiger Lang? Lang was born in the German industrial town of Braunschweig in 1943. As a youngster, he had little interest in studying, instead preferring to learn the trade of watchmaking.

After a short stint at the Hamilton Watch Company, the young journeyman joined Heuer in Biel in mid-1964. Period photographs show Mr. Lang working on the assembly bench for stopwatches, which was Heuer's main business area.

In 1968 Lang was happy to move back to Germany taking to take new responsibilities at the Heuer-IWC shared service center based in Frankfurt.

Just a year later, Heuer launched their first automatic chronograph in a collaboration with Breitling, Buren, and Hamilton.

Gerd-Rüdiger Lang was asked by Jack Heuer to bring a boxful of new watches to France to be featured in the film "Le Mans," starring Hollywood legend Steve McQueen (evolving into a lifelong promotion for the Heuer Monaco). Next morning he began his journey to Le Mans in his MG and breathed in the authentic racing atmosphere, which can be so highly addictive. Gerd-Rüdiger Lang embraced it wholly, discovering his talent as a timekeeperin the Olympic Games and motorsport for the years to come. Heuer split-second chronographs No.63 and No.67 were his personal favorite stopwatches.

Later, in 1974, the Heuer service center moved to Munich. The entire watch industry was hit hard by the quartz crisis during the late 1970s, and Heuer decided to close its German branch in 1980. In the same year Gerd-Rüdiger Lang qualified as a master watchmaker and was invited to take over service operations for Heuer watches in Germany. It was then that he established his own workshop in Munich and became independent.

Gerd-Rüdiger Lang was invited by Jack Heuer to be an official timekeeper during the 1980 Olympic Games in Moscow and was presented with a personalized orange Heuer Senator GMT with Olympia Moscow 1980 logo.

He then went onto found Chronoswiss in 1983 registering the trademark in 1984. Gerd-Rüdiger Lang had a great talent to foresee the future believing in the renaissance in mechanical movements when almost no one else did.

He assumed that a quartz watch was put in the bin after a few years whilst a mechanical timepiece will be put in a museum after 100 years!

Without doubt Gerd-Rüdiger Lang can be seen as one of the first and most successful independent watchmakers. He was a great Mentor to me right up until his passing on March 2nd 2023 in Starnberg, Germany.

He will be greatly remembered.

HEUER
1/10 SEC
CHR.63
HEUER
Москва 1980

BIBLIOGRAFIE/
BIBLIOGRAPHY

Jack Heuer: The Times of My Life. Foreword by Jean-Christophe Babin. TAG Heuer 2013.

Dipl. Ing. Jack W. Heuer: Technik der Handzeitmessung. Technische Rundschau, Bern/Schweiz, Nr14/1961.

Heuer Catalogues and Product Information Leaflets: 1962 to 1990.

Heuer-Leonidas SA: Rapport de Gestion 1972.

Ross Povey - Carolina Porsche: Taste- Makers Revolution Asia 65, UK 28.

Jeannot-Foulkes-Waugh: TAG Heuer Carrera; Ahead of its time, 1963-2018. Portfolio VII by Arno Haslinger, TAG Heuer 2018.

Foulkes-Brunner-Jeannot: Autavia; Story of an icon. TAG Heuer 2017.

Hans Kocher: Die Chronomatic. Ein Chronograph mit automatischem Aufzug.
Neuer Züricher Zeitung, Beilage Technik, 15. Dezember 1969.

Gerd-Rüdiger Lang- Reinhard Meis: Chronographen, Armbanduhren.
Callwey Verlag, München 1992.

Gisbert L. Brunner- Marc Sich: Heuer und TAG Heuer; Die Beherrschung der Zeit. Editions Assouline, Paris 1997.

Bonhams-TAG Heuer: The Haslinger Collection. Wednesday 15 December 2010, London.

Werner Deisenroth- Arno Haslinger: Kalender Time Race Part I and II.

Designers Digest: The culture of Design and Visualisation No 110 and 111.

Car Graphic Edition 626 and 627 Heuer Chronographs and Motorsport, Tokio, Japan.

Heuer 1960 -1982: Summary by TAG Heuer.

Wolfgang Koll: Die Heuer für die Bundeswehr. Uhren Magazin 11/96.

TAG Heuer: Travel Through Time. Globetrotter 15 Days, 10 Cities, 2017.

Materialamt der Luftwaffe: Technisches Handbuch; Armbanduhr mit Doppelstopp-Einrichtung; TKZ 1550 SG, VERS. NR. 6645-12-146-5081.

Siffert in Frankfurt mit Gerd-Rüdiger Lang, Fa. Christ Promotion Uhren für den Sport.

IWC Uhren GmbH Deutschland: Abt. Heuer-Leonidas, Juli 1970.

Nick Foulkes: TAG Heuer: Swiss Avantgarde Since 1860. 150 Years—The Book. Assouline, 2010.

Carrera TAG Heuer: THE RACE NEVER STOPS. Texts: Nicholas Biebuyck, Abrams/ New York 2023

Bildnachweis/
Picture credits

BONHAMS: S. 19
Arno Michael Haslinger: S. 13, 100, 165, 283, 323, 325
Gerd-Rüdiger Lang: S. 151, 323, 325
GP Library Limited / Alamy: S. 132
Paul Maudsley: S. 21
MOTORSPORTARCHIV: S. 4, 32, 33, 63, 106, 123, 156, 157, 309
PORSCHE: S. 24, 114, 117, 122, 123, 150, 186, 224, 227, 232
Laura A.K. Scheybal: S. 211, 213
TAG Heuer: S. 8, 11, 15, 16, 17, 54, 103, 157, 164, 182, 183, 254, 280, 281, 321

DANKSAGUNG/
ACKNOWLEDGMENTS

Dieses Buch ist der Ära von Jack W. Heuer gewidmet, den ich zutiefst bewundere.

Bettina Gottfried – Art Direction
Clemens „The Crown" Kois, NYC – Fotografie
Christiane Resch – Mood- und Grafikunterstützung
Michael Müllner – Uhrmachermeister
Mark Porsche – Für Deine unermüdliche Hilfe
Michael Garbacz – Für die Verwendung von historischen Fotos aus dem österreichischen Motorsportarchiv: www.motorsportarchiv.at

Ein großer Dank gilt all meinen Sammlerfreunden, die Uhren für dieses Buch zur Verfügung gestellt haben: Adrian van Hooydonk, Thomas Bachmann, Joram Scher, Christoph Thurner, Dr. Thomas Wallner und Mario Baroncini.
Ebenso danke ich Jonathan Darracott und Paul Maudsley für ihre Beiträge, ihre Freundschaft und die stetige Unterstützung, die ich in meiner Zeit als Berater beim Auktionshaus Bonhams erfahren durfte.

Mein besonderer Dank gilt Antoine Pin, Nicholas Biebuyck, Jean-Christophe Babin, Frederic Arnault, Stephane Bianchi und allen wunderbaren TAG Heuer-Freunden für ihre wertvolle Unterstützung in den vergangenen 25 Jahren.
Es war eine außergewöhnliche Reise, Teil des Heuer-Collectors-Universums zu sein – geprägt von faszinierenden Produkten, inspiriert von den glorreichen Tagen des Motorsports.

Zum Schluss möchte ich onthedash erwähnen – mit seinem großartigen Heuer-Katalog und den Referenztabellen, zu denen ich in den letzten zwei Jahrzehnten teilweise beitragen durfte.

This book is dedicated to the era of Jack W. Heuer, whom I admire immensely.

Bettina Gottfried, Art Direction
Clemens "The Crown" Kois, NYC; Photography
Christiane Resch; Moods and graphic support.
Michael Müllner; Watchmaking Champion.
Mark Porsche; For your tireless help.
Michael Garbacz; For usage of period photos from the Austrian Motorsport Archive: www.motorsportarchiv.at

Many thanks to all collector friends who provided watches to be featured in the book: Adrian van Hooydonk, Thomas Bachmann, Joram Scher, Christoph Thurner, Dr. Thomas Wallner, and Mario Baroncini.
Jonathan Darracott and Paul Maudsley for your contribution, friendship, and the continuous support I have experienced as a consultant at Bonhams auction house.

I am grateful to Antoine Pin, Nicholas Biebuyck, Jean-Christophe Babin, Frederic Arnault, Stephane Bianchi, and the all lovely friends of TAG Heuer for your immensely important support during the last 25 years.
It has been an amazing journey to have the opportunity to be part of the Heuer collectors' universe indulged by amazing products inspired by the heyday of motorsport.

I would like to mention onthedash, with great Heuer catalog and reference tables to which I have occasionally been allowed to contribute over the last two decades.

CALLWEY

Bücher machen glücklich

seit

1884

Seit seiner Gründung hat sich der Callwey Verlag als einer der führenden deutschen Verlage etabliert, bekannt für seine hochwertigen Publikationen in den Bereichen Architektur, Design, Garten, Wohnen und Lifestyle. Was vor mehr als 140 Jahren mit der Gründung einer Verlagsbuchhandlung durch Georg D.W. Callwey begann, entwickelte sich schnell zu einem renommierten Verlagshaus, das Generationen von Lesern inspiriert.
Die inzwischen vierte Generation führt das Erbe mit derselben Hingabe und demselben Innovationsgeist fort, die schon die Gründer angetrieben haben.

Der Callwey Verlag steht für Inspiration, Kreativität und höchste Qualität. Jedes Buch, das das Verlagshaus verlässt, ist ein Meisterwerk für sich, oft über Jahre hinweg geplant und entwickelt, sorgfältig kuratiert, gestaltet und mit den besten Papieren liebevoll verarbeitet – seit 1884 hat sich hieran nichts verändert.

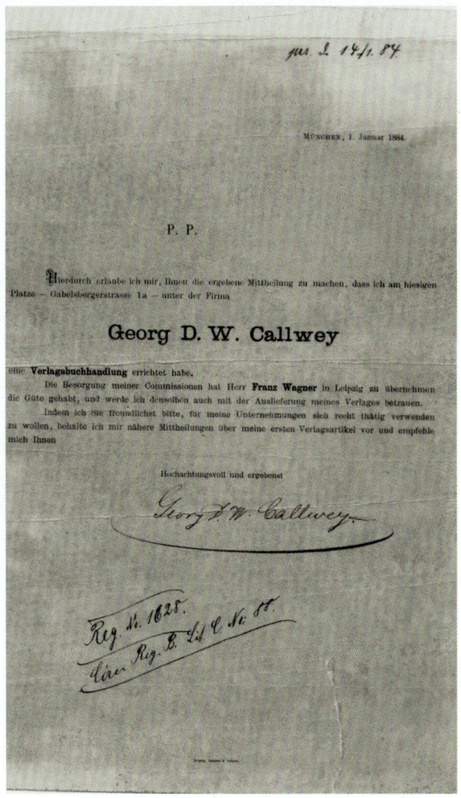

München, 1. Januar 1884.

P. P.

Hierdurch erlaube ich mir, Ihnen die ergebene Mittheilung zu machen, dass ich am hiesigen Platze – Gabelsbergerstrasse 1a – unter der Firma

Georg D. W. Callwey

eine **Verlagsbuchhandlung** errichtet habe.

Die Besorgung meiner Commissionen hat Herr **Franz Wagner** in Leipzig zu übernehmen die Güte gehabt, und werde ich denselben auch mit der Auslieferung meines Verlages betrauen.

Indem ich Sie freundlichst bitte, für meine Unternehmungen sich recht thätig verwenden zu wollen, behalte ich mir nähere Mittheilungen über meine ersten Verlagsartikel vor und empfehle mich Ihnen

Hochachtungsvoll und ergebenst

Georg D. W. Callwey

Eine Reise durch die Jahrhunderte: Die historischen Verlagsräume, Mitarbeiterinnen und Mitarbeiter des Verlags in den 20er und 60er Jahren, der Verlagsgründer Georg D.W. Callwey, Mitglieder der Verlegerfamilie, darunter Marcella Prior-Callwey.

IMPRESSUM

CALLWEY1884

©2025 Callwey GmbH
Klenzestraße 36 · 80469 München · buch@callwey.de
Tel. / Fon: +49 89 8905080-0 · www.callwey.de

Wir sehen uns auf Instagram: www.instagram.com/callwey
Follow us on Instagram: www.instagram.com/callwey

ISBN 978-3-7667-2798-5 · 1. Auflage 2025
ISBN 978-3-7667-2798-5 · 1st edition 2025

Bibliografische Information der Deutschen Nationalbibliothek: Die Deutsche Nationalbibliothek verzeichnet diese Publikation in der Deutschen Nationalbibliografie; detaillierte bibliografische Daten sind im Internet über dnb.dnb.de abrufbar.

Bibliographic information of the German National Library: The German National Library lists this publication in the German National Bibliography; detailed bibliographic data is available online at dnb.dnb.de.

Viel Freude mit diesem Buch wünschen Ihnen:
We hope you enjoy this book:
Projektleitung / *Project Management*:
MAJA FANGMANN
Lektorat / *Proof reading*: JUDITH MARNET
Schlusskorrektur / *Final Proofreading*:
DEUTSCH / GERMAN: JUDITH MARNET
ENGLISCH / ENGLISH:
ALISON MOFFAT-MCLYNN
Gestaltung & Satz / *Design & Layout*:
DANIELA PETRINI
Herstellung / *Production*: OLIVER MEIER
Art Direction / *Art Direction*: MARIO LOMBARDO

Autor/*Author*:
Arno Michael Haslinger ist Experte für historische Uhren und Autos sowie leidenschaftlicher Motorsport-Enthusiast. Seit mehr als 25 Jahren widmet er sich der Forschung, Dokumentation und dem Sammeln klassischer Zeitmesser, insbesondere der legendären Heuer-Chronographen.
Nach seinem Abschluss an der Wirtschaftsuniversität Wien führte ihn seine internationale berufliche Laufbahn in die Luxus- und Automobilbranche mit Stationen in Frankfurt, Paris und London.
2011 gründete er seine Beratungsfirma TimeRace Consulting, berät Bonhams Watches und ist Mitglied der Jury des Grand Prix d'Horlogerie de Genève (GPHG). Seine Bücher gelten unter Sammlern, Uhrenliebhabern und Motorsportfreunden als Standardwerke zur Geschichte der Marke Heuer.

Arno Michael Haslinger is an expert in historical watches and vintage cars, and a passionate motorsport enthusiast. For more than 25 years he has focused on researching, documenting, and collecting classic timepieces, with a particular passion for the legendary Heuer chronographs.
After graduating from Vienna University of Economics and Business, Haslinger pursued an international career in the luxury and automotive sectors, holding top positions in Frankfurt, Paris, and London.
He founded TimeRace Consulting in 2011, serves as a consultant for Bonhams Watches, and is a member of the jury for the Grand Prix d'Horlogerie de Genève (GPHG).
His books are widely regarded as authoritative references on the history of the Heuer brand among collectors, watch enthusiasts, and motorsport fans alike.

Hinweis: Uns ist es ein Anliegen, dass sich alle Geschlechter wahrgenommen und wertgeschätzt fühlen. Im Sinne einer besseren Lesbarkeit der Texte verzichten wir jedoch auf die gleichzeitige Verwendung der Sprachformen männlich, weiblich und divers (m/w/d). Wo dies möglich ist, bemühen wir uns darum, alle Formen miteinzubeziehen, oder um neutrale Formulierungen. Sämtliche Personenbezeichnungen gelten gleichermaßen für alle Geschlechter.